99 New Discoveries in Astronomy

99 New Discoveries in Astronomy

P.J. Tomlin

AuthorHouse™
1663 Liberty Drive
Bloomington, IN 47403
www.authorhouse.com
Phone: 1-800-839-8640

© 2012 by P.J.Tomlin. All rights reserved.

No part of this book may be reproduced, stored in a retrieval system, or transmitted by any means without the written permission of the author.

Published by AuthorHouse 01/11/2013

ISBN: 978-1-4772-3511-9 (sc)
ISBN: 978-1-4772-3512-6 (hc)
ISBN: 978-1-4772-3513-3 (e)

Any people depicted in stock imagery provided by Thinkstock are models, and such images are being used for illustrative purposes only.
Certain stock imagery © Thinkstock.

This book is printed on acid-free paper.

Because of the dynamic nature of the Internet, any web addresses or links contained in this book may have changed since publication and may no longer be valid. The views expressed in this work are solely those of the author and do not necessarily reflect the views of the publisher, and the publisher hereby disclaims any responsibility for them.

CONTENTS

Acknowledgements...i

Preface..iii

Chapter One	The Hubble Constant Enigma 1
Chapter Two	A Question of Time 18
Chapter Three	Time, Mass, and the Age of the Universe 24
Chapter Four	The Sun 37
Chapter Five	Gravity and Time 53
Chapter Six	The Earth and the Moon... 69
Chapter Seven	Jupiter and Saturn.. 90
Chapter Eight	Chaos, Destruction, and Regeneration: A Revised Cosmology... 106
Chapter Nine	Lists of Findings and Proofs 129

ACKNOWLEDGEMENTS

This work would not have been possible but for the findings of Professor Sol Perlmutter and his colleagues on distant supernovas, which provided the key data about the constancy of the Hubble constant up to the limit of their observational power. Figure 1.1 in chapter 1 is based on calculations of their observations and is a testimony to the accuracy and precision of their work.

PREFACE

This book is intended not only for the reader interested in science and particularly interested in astronomy but also for the professional astronomer and professional physicist. To facilitate this, each chapter has been divided into two. The first part describes the findings and the scientific principles underlying those findings and assumes no knowledge of mathematics beyond that of everyday usage. The appendix to each chapter gives the technical details that support the new findings.

This work started because of curiosity about the mysterious Hubble constant. This is the constant which shows that the universe is expanding and that it is doing so in a constant fashion so that the velocity of any receding galaxy is exactly proportional to its distance from Earth. There is a mystery about this constant, not only about the physics which underlie it but also that it is paradoxical. Under standard Euclidean geometry the constant should not be constant, yet detailed analysis of data from the most far-flung supernovas show that it is precisely constant, apart from the effects of relativity, up to the limit of accurate measurements of data from such supernovas.

Eventually a solution was found. It involved time but appeared too outrageous to suggest that it was a mathematical fix with no application to reality. Yet the same solution resolved three other paradoxes which have plagued astronomy for years. It also provided a perfect demonstration of Einstein's special relativity theory. A search was made through the major branches of physics, looking for any contradiction within established and proven physics—with the emphasis on proven. None were found, but what was noticed was that in various branches of physics, equations existed which showed that time and mass were inversely related. One such equation goes back as far as Galileo's time. It was this equation which led to the great outburst of new discoveries that are listed in chapter 9 of this book.

Some discoveries were negative. The hypotheses about dark matter and dark energy had to go, as did the Big Bang theory. One discovery explained why, despite over fifty years of intensive effort, scientists around the world have failed to obtain surplus energy from hydrogen fusion. They never will. It emerged that atomic fusion absorbs energy whilst atomic fission releases energy. The greatest energy release in the universe, one that blows giant stars to pieces, is helium fission, but it requires a very high temperature. On the contrary, endothermic hydrogen fusion was found to play a vital role in maintaining the thermal stability of the sun and indeed of all stars. Solar energy arises not from hydrogen fusion but a mechanism involving the behaviour of time.

But there were other surprises. One discovery explained the mechanism of the most spectacular sight in the solar system, the monstrously large but magnificent ice fountains found on Saturn's baby moon, Enceladus. These fountains, if they existed on the Tibetan plateau, would be higher than Mount Everest. Other surprises were the emergence of why the Earth has a slight wobble, why the Pacific Ocean is so deep, why "Snowball Earth" occurred, and why plate tectonics did not start until well after the emergence of mammals. Other surprises were why the surface of Saturn is so relatively warm and why the interior of Jupiter is so hot.

On a more academic level, the mode of action of gravity was found, and the gravitational energy output of any mass could be calculated. More significantly, a fifth fundamental force of nature, one that had been previously forecast by one of the world's greatest theoretical atomic physicists, was found and quantified, and its mode of action identified.

This book will challenge much established thinking about astronomy, just as Charles Darwin's book *On the Origin of Species* challenged established thinking about biology. But what a challenge!

Preface v

Questions, Questions, Questions

What drives the expansion of the universe? What controls the energy for the expansion?

Why is solar energy output so very stable? What is its source and how is it controlled?

Why have scientists, despite fifty years of effort, failed to obtain surplus energy from hydrogen fusion? Was the H bomb test a lucky accident?

What is the greatest explosive force in the universe?

What happened after the greatest catastrophe ever struck the Earth? What causes plate tectonics and so earthquakes? Why were they so late in developing?

Do dark matter and dark energy exist? Why are the Pioneer space probes accelerating?

Where does the energy of the background microwave radiation come from?

Is there a fifth fundamental force of nature? What does it do?

How does gravity act as an attractive force? Can gravitational energy be quantified?

What is the greatest dynamic spectacle in the solar system? How does it work?

How did Saturn's ring system develop?

How is the Hubble constant (the velocity distance ratio of receding galaxies) able to defy normal geometry?

How old is the universe and where does time fit in all this? Is its behaviour the ultimate controller of the universe?

Preface vi

The answers to all these and many other questions emerged as the discoveries gradually unfolded. Just as the Rosetta stone provided the key to solving the mysteries of the hieroglyphics in the Egyptian pyramids so the key to these discoveries lay in solving the triad of the Hubble constant enigmas, (the dimensional description of the Hubble constant, why and how does it contradict Euclidean geometry, what are the physics underlying the constant). Yet it is all based on *proven* science.

CHAPTER ONE

The Hubble Constant Enigma

Some eighty years ago, the doyen of American astronomy, Edwin Hubble, made a discovery which revolutionised our understanding of astronomy. He found that our galaxy is surrounded by other galaxies which are receding from us. Further, the more distant a galaxy is, the faster is its recessional velocity. He found that there was a fixed or constant relationship between the velocity of the receding galaxy and its distance. This ratio has become known as the Hubble constant. Certain consequences followed. The first was the realisation that the universe is expanding. It was appreciated that tracing backwards, the universe must have had a singular point of origin, and that it also allowed the calculation of the age of the universe and so the development of the Big Bang theory. It also meant that the universe is spherical in shape. The orderliness of the expansion meant that there must be some overarching mechanism that is controlling the expansion. There were other unknowns: (a) Is the constant a local phenomenon or does it extend to the edge of the visible universe? (b) Mathematical analysis suggests that it should not be constant, that is, there must be a mechanism that keeps it constant, and if that is so, what is that mechanism? (c) The constant merely describes a ratio, so what are the underlying physics behind the expansion of the universe? (d) What exactly is the constant defining?

Hubble's estimate of distances, although state of the art at the time, was off by a factor of 100. But this does not disturb the ratio. Soon afterwards came the discovery of variable stars. These are stars whose brightness varies in a regular rhythmical way, and the longer the cycle, the greater the peak brightness. There are a number of such stars within our galaxy that are sufficiently near that the distances could be ascertained by conventional means. Thus if the period of the cycle was known and the peak brightness measured, a reasonably accurate estimate of the distance could be calculated, based on the inverse square law (double the distance and the brightness decreases fourfold, and so on).

Similar variable stars could be seen in our neighbouring galaxy, Andromeda. Although the distance made the starlight dimmer, calculating the distance using a variety of different variable stars within Andromeda gave near identical results. The search was then on for other nearby galaxies in which variable stars could be detected. This confirmed that the Hubble constant was around 50 km per second per mega parsec (Mpc; a parsec is 3.26 light years). Based on these figures, the best estimate of the age of the universe was 13.7 billion years.

At the time, the brightness of a distant star was a subjective estimate made by a team of trained scientists examining photographic plates. That is, the estimate was subject to observer error. It was not realised that the photographic emulsion consumed a small amount of light energy to release the silver in the emulsion. This is constant per molecule of emulsion. Brightness is measured using a logarithmic scale. The effect of this is if the incoming light is very bright, this energy loss is insignificant. But when the incoming light was very dim, proportionately less energy was available for the brightness to appear on the photographic plate. This has the effect of underestimating the brightness of the dim distant variable stars and so led to an overestimate of the distance and an underestimate of the value of the Hubble constant. Modern methods now use very sensitive and very accurate light detectors and have shown that the Hubble constant is approximately 52 km/sec/Mpc, yielding an age of the universe of around 12.53 billion years.

The Extent of the Constancy of the Hubble Constant

To test whether the constant extends to the edge of the visible universe, two sets of published data of Type 1A supernovas were used (Hamuy et al., 1996; Perlmutter et al. 1999). Type 1A supernovas are supernovas of exploding neutron stars (that is, highly compressed helium atoms, where their encircling orbiting electrons are either tightly compressed around the helium nuclei or forced into the nuclei). Helium fission is the most powerful transient force in nature; it is responsible for all supernovas. Type 1A supernovas have little to no hydrogen in their spectra. Neutron stars have a limited range of mass, below 8 solar masses.(A solar mass is the mass of our Sun and equals very close to 2×10^{30} kg.)Any mass greater than this with the density of a neutron star or higher will have the

Chapter One 3

gravitational strength of a black hole. Neutron stars gradually lose mass, as energy, as they slowly heat up. (See later.)

It had been observed that Type 1A supernovas emit a set amount of light that decays in a particular fashion over a particular time. They therefore can act as standard candles. It follows that with uniform expansion of the universe, all such supernovas with the same velocity should have the same magnitude, as they will be at the same distance from us. But supernovas can occur anywhere in a galaxy, including behind dust clouds, which can obscure some of the light. The results (see Tables 1.A1 and 1.A2 in this chapter's appendix) show that this is the case. Variations up to 0.5 of magnitude have occurred between galaxies of similar velocities.

In the calculation of distances, the observed brightness or magnitude must be corrected by three factors, derived from the template of the light decay curve to obtain the peak effective brightness. They are all influenced by time, as is the observed brightness.

The energy released by a Type 1A supernova is prodigious. At a distance of 10 parsecs (32.6 light years), such a supernova would be over 200 times brighter than the brightness we now perceive from the sun, which is only a little more than 8 light minutes away. If the sun were to become a Type 1A supernova, Earth would be vaporised.

The first set of data, Calan/Tololo set, consisted of eighteen supernovas that were at distances of between 68.2 and 597 Mpc. The second set were from the Supernova Cosmology Project (SCP) and consisted of forty-two Type 1A supernovas whose distances ranged from 896 to 3728 Mpc, or slightly more than 12 billion light years, (a light year is the distance traversed by a photon travelling at the speed of light, for one year and equals approximately 9.5 thousand billion kilometres) The Calan/Tololo data yielded a value of 51.8 +/-1.6 km/sec/Mpc for the Hubble constant.

Calculating the Hubble constant from the SCP data proved more troublesome until it was realised that they were receding sufficiently fast that relativistic time dilatation was skewing the results. The distances were calculated from the magnitude or brightness of the Type 1A supernovas. These in turn follow a precise time sequence. If, through special relativity, time for these supernovas was being slowed, this would reduce the rate of photon emission. That is, at high velocities the magnitude or brightness would be dimmed because of this relativity effect. This would affect four parameters used to calculate the effective magnitude. The extent of the time dilatation could be calculated from the velocity using the standard special relativity time velocity equation. To a lesser degree, special relativity's time slowing would also affect the red shift, used to determine the velocity, and again this could be allowed for. When this was done the value for the Hubble constant from the SCP data was 52.0 +/-1.3 km/sec/Mpc. There was no statistical difference between the two sets of data. The results are displayed in Figure 1.1.

The conclusion must be that after allowing for the relativity effect the Hubble constant is constant to the edge of the observable universe, and it has a value of 52.0 +/-1.3 km/sec/Mpc. An Internet search shows that there is variety of values for the Hubble constant, though for the more modern measurements, they are all in the low fifties. The samples of galaxies observed were all relatively small. There is one exception. In Gribben's book, *The Birth of Time,* from an analysis of the published data of 1,388 galaxies within a range of 100 Mpc, he found that the Hubble constant was 52 +/-6 km/sec/Mpc. His error range from such a large sample suggests that there was considerable variation or error in his sample, which happily cancelled out. Mathematically, one would expect this from such a large sample.

The Mathematical Puzzle

Calculation using simple Euclidean geometry shows that the Hubble constant should *not* be constant. The actual equation (see this chapter's appendix for details) is:

Equation 1.1

Hubble constant (Ho) = velocity/distance = $652/(\text{age} - D_T)$

where age is the age of the universe (in Earth time—that is, time as assessed on Earth) and D_T is the time taken for light to travel the distance.

Chapter One

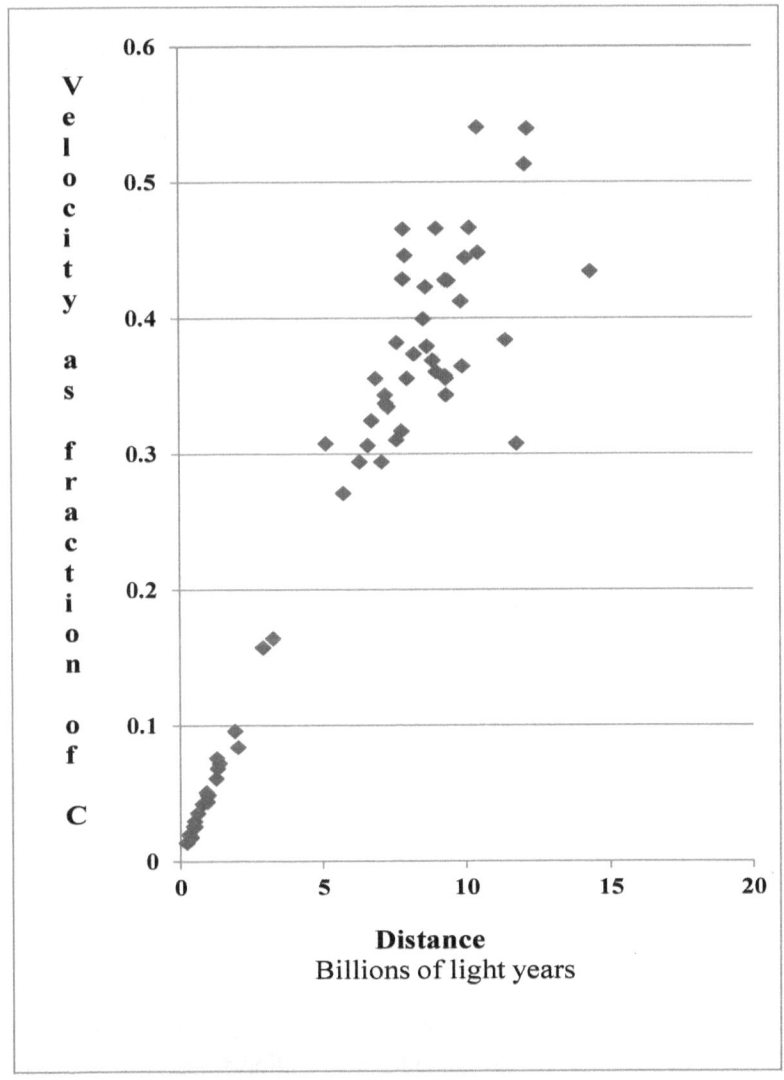

Figure 1, The relationship between distance and velocity for 60 supernovas type Sn1a (there is some overlap in the data points) after correcting for the effects of special relativity. The linear relationship shows the constancy of the Hubble constant.

Attention is drawn to the denominator (age-D_T), which describes the age of the universe at the time of the supernova explosion. This period of time will vary with different supernovas, and hence the Hubble value should not be constant, but it is! The only mathematical mechanism to ensure that the value is constant is to multiply the 652 in Equation 1.1 by the ratio of the age of the universe at the time of the supernova explosion to the present age (using Earth time). This ratio has a profound significance, which will be examined in detail in chapter 2; sufficient to say that the only way the Hubble constant can be constant up to near the edge of the universe is if time itself is and has always been slowing. The universe uses its own system of time, that is, cosmological time, in which time is slowing exponentially in contrast to Earth time, which rests on the (unproven) assumption that the pace of time, the period of the second, has been unchanging since the dawn of time.

What Is the Hubble Constant in Terms of Physics?

The Hubble constant is velocity divided by distance, that is, distance divided by time multiplied by distance. Dimensionally, this gives a number divided by time. This describes a frequency, which is clearly inappropriate. But it shows that the velocity of a galaxy increases with distance. That is, it is describing an acceleration, in which case dimensionally, it should be velocity divided by time. But clearly distance is involved. The only fixed relationship between distance and time is c, the velocity of light. It follows that the time must be the time for light to travel that mega parsec distance. Converting a Hubble constant of 52 km/sec/Mpc into acceleration units results in an acceleration constant, with a value of 5.05×10^{-10} m/sec/sec. This acceleration, if maintained for 10 billion years, will result in a velocity approximately half the velocity of light. This implies that the life expectancy of a galaxy, including its formative years, is slightly less than 20 billion years, as at the speed of light mass ceases to exist. It has been shed as energy. Alas, it is not that simple. Special relativity also comes into play, slowing down the internal time of the fast moving galaxy.

There is a curious coincidence of figures, only it is not a coincidence. The two Pioneer probes are accelerating out of the solar system. This acceleration is around 5×10^{-10} m/sec. There has been no explanation for this acceleration. The outer stars of spiral galaxies have been found to have an acceleration of around 5×10^{-10} m/sec. This has been variously attributed to change in Newton's law of gravity, when the gravity is extremely low, or to unknown or dark matter within the galaxy—with such dark matter being 8 to 9 times the mass of the galaxy. How so much matter could be distributed within the galaxy without disturbing the central geometry of the galaxy has not been explained. Similarly, black holes within galaxies move with the same acceleration as the galaxies themselves, as otherwise the geometrical structure of the spiral galaxies could not be maintained. Galaxies are accelerating at 5.05×10^{-10} m/sec. This includes the dust within a galaxy. All this has been attributed to dark energy. If the energy arose outside the galaxy, the geometrical structure and cohesion of the spiral galaxies could not be maintained.

There are five different types of masses involved: dust, the Pioneer probes, the outer stars of a galaxy, black holes, and whole galaxies, between them varying in mass by thousands of orders of magnitude yet all showing the same acceleration point to a fundamental force in nature. This force has some similarities to gravity, in that gravity too produces the same acceleration irrespective of the mass being gravitationally accelerated. Such a force eliminates the necessity of dark energy and dark matter. That is, there must be some other mechanism, one that arises from within the mass being accelerated. Without this acceleration force, the universe would have collapsed into a gravitational heap of matter billions of years ago.

A Fifth Fundamental Force of Nature

Many years ago the distinguished atomic physicist, Burkhard Heim, proposed that in addition to the standard four fundamental forces of nature, the strong and weak intranuclear forces, electromagnetic force, and gravity, there must exist two additional forces. He called them interactions. Heim was a formidable theorist. Despite not having the use of his hands (they were destroyed in an accidental explosion), he was able to calculate the masses of a number of subatomic particles, which later experimentation subsequently confirmed. One such force has tentatively been identified in the laboratory and is called the gravito-magnetic force. The other, he suggested, has some similarities with gravity but is antithetical to gravity. The key similarities are that both act on mass in a manner which is independent of the amount or size of the mass. They both produce acceleration. That is, they both must act on the individual nucleons with the mass. They both arise from nucleons.

The Hubble acceleration force fits closely with Heim's proposed fundamental force. The force must arise from all the surfaces of the individual nucleons but, unlike gravity, must have a very short range or perhaps no range at all. It just acts on the surface of the nucleon. But it is a compressive force holding the nucleon's quarks together. If the nucleon is stationary, the compressive force is the same across the whole surface. Forces opposite each other around the surface of the nucleon then cancel each other so that there is no movement. If the nucleon is moving, the prow of the nucleon overtakes the force at that point, so its compressive action at that point is lost. At the stern of the moving nucleon, the force is unopposed, resulting in acceleration in the direction of the movement.

The force does work. That is, it consumes energy. If the mass is very large with a huge quantity of nucleons, the energy used must be very substantial. Accelerating a black hole that has a mass of a million suns, even if the acceleration was 5×10^{-10} m/sec/sec, would require 1.25×10^{18} joules of energy, the equivalent of approximately 1125 kg of mass every second. Such an amount of energy can only come from the mass itself. The same applies to gravity, but the gravitational constant is approximately 10 times weaker than the acceleration constant—although there may well be many more gravitons promoting gravitational acceleration. The conclusion must be that nucleons are very, very slowly losing some of their substance to provide the energy for the expansion of the universe and for the gravitational energy released. This is a continuous process and is controlled by the expansion of time (the relationship between time and mass is discussed in the next chapter).

An Unexpected Finding

Black holes are accelerated to the same extent as intra-galactic dust. Acceleration depends upon time. Einstein's special theory of relativity says that for a moving particle around a black hole, time is slowed so that the

Chapter One

second is infinitely long, all because of the immense gravity of the black hole. If time were infinitely prolonged around and within a black hole, there could not be the uniform acceleration of the various components of a galaxy. It follows therefore that time around and within the black hole is normal. It also follows that the time slowing projected by relativity theory applies only to the internal time of an object that is moving through the black hole's gravitational field. In the black hole, the nucleons must be stationary relative to each other and so stationary within the internal gravitational field. This would apply even if the black hole was revolving. This casts doubt on whether the curvature of space-time actually exists; instead, the curvature deduced is a mathematical consequence of the time slowing for the object traversing the gravitational field being accentuated by the gravitational pull. This in turn is a consequence of the rate of the arrival of gravitons on the moving object.

It also follows that within the black hole, nucleons maintain their integrity, so that each nucleon puts out the same amount of acceleration energy. The same argument applies to the release of gravitational energy.

APPENDIX 1

To verify that the Hubble constant extends to the limits of the universe, the relevant published data of two populations of Type 1A supernovas were taken (Perlmutter et al., *Astronomical Journal* 2, 1999, 570-585). These were the Calan/Tololo data and the SCP SN1a data. Supernovas act as standard candles whose light declines in a very specific way. The Hubble constant is the ratio of velocity to distance of any galaxy. Distance was calculated from the magnitude of the supernova, using a derivation of the equation linking magnitude to distance. The original equation was:

Equation Appendix1.1

$$\text{Log distance} = (M + 19.7 - k - 25)/5$$

The equation is based on the inverse square law and relies on the calculation that a Type 1A supernova at a distance of 10 parsecs would have a magnitude of -19.7. M is the magnitude of the supernova in question. The distance is in mega parsecs. More recent work improved this equation by substituting the effective magnitude for the observed peak magnitude using various corrections (see Perlmutter's paper). These corrections are all time sensitive.

Equation Appendix 1.2.

$$\text{Effective mag} = \text{Observed mag} - Ax - K - s$$

Velocity was taken from the z shift for each supernova, and in Tables 1.A1 and 1.A2, the velocity is also given as a fraction of the velocity of light. For the Calan/Tololo data, Table 1.A1, which consisted of eighteen supernovas that were in middle distance to the edge of the visible universe, the calculations showed that the Hubble constant had a value of 51.8 +/-1.35 km/sec/Mpc. This result is in line with the results from small sample surveys to be found on the Internet.

The SCP data referred to supernovas that were much farther distant and had very high velocities. The Ho calculations gave anomalous results until it was realised that the data were being affected by relativity's time dilatation. For the calculation of distance, each of the four components used to determine distance needed correcting using special relativity's time velocity equation. Since these components were in log format, this meant using a corrective, where vc is the velocity expressed as a fraction of the velocity of light.

Equation Appendix 1.3.

$$\text{Relativity factor} = \log(1/(1-vc^2))^{0.5}$$

Magnitude is measured in base 4 logarithm. Multiplying the expression in Equation 1.A3 by 2.5 allows for this. Since there are four time dependent components, the resulting equation for determining distance became:

Equation Appendix 1.4.

$$\text{Log distance (Mpc)} = (M + 19.7 - Ax - K - s - (10 \times \text{relativity factor}) - 25)/5$$

The velocity was derived from the red shift. That in turn depends upon the time frame or pace of time of the moving body. Apart from the universal slowing of time (see later), the time frame is additionally affected by relativity's time change. Therefore the correction is to multiply the velocity by the antilog of the relativity factor. This will give what the value of the Hubble constant would be if it were not reduced in value by the relativistic reduction in the velocity (and so the frequencies of the emitted light waves in the light spectra) of the fast moving supernovas. When this was done, the Hubble constant was shown (Table 1.A2) to be 52 +/-1.2 km/sec/Mpc. To a degree this is somewhat artificial, but it shows that without this interference due to relativistic time dilatation, the Hubble constant would be truly constant to the limit of accurate observation.

Putting the two sets of data together, it is clear that within the limits of accuracy of measurement, the Hubble constant is constant until the velocities are so high that relativity effects come into play. That is, the rate of the expansion of the universe very near the edge of the universe slows down. The Hubble constant is indeed constant relative to the time frame of the individual galaxies throughout the universe. In this, it shows that the force responsible for the acceleration of the galaxies conforms with the criteria that define a fundamental force of nature.

Notes on Tables 1.A1 and 1.A2

Column 1	The reported Z value of the supernova.
Column 2	The velocity of the supernova expressed as a fraction of the velocity of light.
Column 3	M, the reported effective peak magnitude of the supernova.
Column 4	Distance (Mpc) derived from the reported peak magnitude.
Column 5	The apparent Hubble factor.
Column 6	The relativity correction factor from Einstein's $T = t(1/(1-V^2)^{0.5}$.
Column 7	Relativity corrected peak magnitude from column 3, i.e. 10(log column 6).
Column 8	Distance (Mpc) derived from Equation 1.A4 after correcting for relativity effects.
Column 9	Hubble constant corrected for relativity's distortion effect on distance.
Column 10	The Hubble constant after correcting for the effects of relativity on velocity.

Table Appendix 1 The Calan/Tololo Data

	Supernova	z [1]	v [2]	Peak M [3]	Distance [4]	Ho [5]	Rel factor [6]	Peak M [7]	Distance [8]	Ho [9]
1	1992al	0.014	0.014	14.47	68.23	61.1	1.00010	14.47042	68.2	61.1
2	1992bo	0.018	0.018	15.61	115.35	46.4	1.00016	15.61069	115.4	46.4
3	1992bc	0.02	0.020	15.18	94.62	62.8	1.00020	15.18085	94.7	62.8
4	1992P	0.026	0.026	16.08	143.22	53.8	1.00033	16.08143	143.3	53.7
5	1992ag	0.026	0.026	16.28	157.04	49.0	1.00033	16.28143	157.1	49.0
6	1990O	0.03	0.030	16.26	155.60	57.0	1.00044	16.2619	155.7	56.9
7	1992bg	0.036	0.035	16.66	187.07	56.7	1.00063	16.66272	187.3	56.7
8	1992bl	0.043	0.042	17.19	238.78	52.9	1.00089	17.19385	239.2	52.8
9	1992bh	0.045	0.044	17.61	289.73	45.5	1.00097	17.61421	290.3	45.5
10	1990af	0.05	0.049	17.63	292.42	50.0	1.00119	17.63517	293.1	50.0
11	1993ag	0.05	0.049	17.69	300.61	48.7	1.00119	17.69517	301.3	48.6
12	1993O	0.052	0.051	17.54	280.54	54.2	1.00129	17.54558	281.3	54.1
13	1992bs	0.063	0.061	18.24	387.26	47.3	1.00187	18.2481	388.7	47.2
14	1993B	0.071	0.068	18.33	403.65	50.9	1.00235	18.34021	405.5	50.8
15	1992ae	0.075	0.072	18.43	422.67	51.2	1.00262	18.44135	424.9	51.1
16	1992bp	0.079	0.076	18.27	392.64	58.0	1.00289	18.28254	394.9	57.8
17	1992br	0.088	0.084	19.28	625.17	40.4	1.00356	19.29543	629.6	40.2
18	1992aq	0.101	0.096	19.16	591.56	48.6	1.00463	19.18007	597.1	48.4
					Mean	51.9			Mean	51.8

Table Appendix 1.2 The SCP SN1a Data

	Supernova	z [1]	v [2]	Effective peak [3]	Distance Mpc [4]	Ho [5]	Relativity factor [6]	Relativity corrected peak [7]	Distance Mpc [8]	Ho [9]	Red shift corrected Ho [10]
1	1997I	0.172	0.157	20.17	942	50.1	1.0126	20.06	896	53.4	54.0
2	1997N	0.18	0.164	20.43	1062	46.3	1.0137	20.31	1005	49.6	50.3
3	1997ac	0.32	0.271	21.86	2051	39.6	1.0388	21.53	1762	47.9	49.7
4	1994F	0.354	0.294	22.38	2606	33.9	1.0463	21.99	2175	42.4	44.4
5	1994am	0.372	0.306	22.26	2466	37.2	1.0504	21.83	2025	47.6	50.0
6	1994H	0.374	0.307	21.72	1923	48.0	1.0509	21.29	1577	61.5	64.6
7	1997O	0.374	0.307	23.52	4406	20.9	1.0509	23.09	3612	26.8	28.2
8	1994an	0.378	0.310	22.58	2858	32.6	1.0518	22.14	2334	41.9	44.1
9	1995ba	0.388	0.317	22.65	2951	32.2	1.0542	22.19	2389	41.9	44.2
10	1995aw	0.4	0.324	22.36	2582	37.7	1.0571	21.88	2068	49.7	52.6
11	1997am	0.416	0.334	22.57	2844	35.3	1.0611	22.05	2244	47.5	50.4
12	1994al	0.42	0.337	22.55	2818	35.9	1.0621	22.03	2215	48.5	51.5

13	1994G	0.354	0.294	22.13	2323	38.0	1.0463	21.74	1938	47.6	49.8
14	1997Q	0.43	0.343	22.57	2844	36.2	1.0647	22.03	2214	49.5	52.7
15	1996cn	0.43	0.343	23.13	3681	28.0	1.0647	22.59	2865	38.3	40.7
16	1997ai	0.45	0.355	22.83	3206	33.2	1.0698	22.24	2448	46.6	49.9
17	1995az	0.45	0.355	22.51	2767	38.5	1.0698	21.92	2112	54.0	57.8
18	1996cm	0.45	0.355	23.17	3750	28.4	1.0698	22.58	2862	39.8	42.6
19	1995aq	0.453	0.357	23.17	3750	28.6	1.0706	22.58	2854	40.2	43.0
20	1992bi	0.458	0.360	23.11	3648	29.6	1.0719	22.51	2763	41.9	44.9
21	1995ar	0.465	0.364	23.33	4036	27.1	1.0738	22.71	3036	38.7	41.5
22	1997P	0.472	0.368	23.11	3648	30.3	1.0757	22.48	2724	43.6	46.9
23	1995ay	0.48	0.373	22.96	3404	32.9	1.0778	22.31	2522	47.8	51.6
24	1996cg	0.49	0.379	23.1	3631	31.3	1.0806	22.43	2663	46.1	49.8
25	1996ci	0.495	0.382	22.83	3206	35.7	1.0819	22.15	2340	53.0	57.3
26	1995as	0.498	0.383	23.71	4808	23.9	1.0828	23.02	3498	35.6	38.6
27	1997H	0.526	0.399	23.15	3715	32.2	1.0907	22.40	2626	49.7	54.2
28	1997L	0.55	0.412	23.51	4385	28.2	1.0976	22.70	3022	44.9	49.3
29	1996cf	0.57	0.423	23.27	3926	32.3	1.1035	22.41	2648	52.8	58.3
30	1997af	0.579	0.427	23.48	4325	29.6	1.1062	22.60	2889	49.1	54.3
31	1997F	0.58	0.428	23.46	4285	30.0	1.1065	22.58	2859	49.7	55.0
32	1997aj	0.581	0.428	23.09	3614	35.6	1.1068	22.21	2409	59.1	65.4
33	1997K	0.592	0.434	24.42	6668	19.5	1.1101	23.51	4391	32.9	36.5
34	1997S	0.612	0.444	23.69	4764	28.0	1.1162	22.74	3070	48.5	54.1
35	1995ax	0.615	0.446	23.19	3784	35.3	1.1171	22.23	2430	61.5	68.7
36	1997J	0.619	0.448	23.8	5012	26.8	1.1183	22.83	3204	46.9	52.4
37	1995at	0.655	0.465	23.27	3926	35.5	1.1296	22.21	2411	65.4	73.8
38	1996ck	0.656	0.466	23.57	4508	31.0	1.1299	22.51	2766	57.1	64.5
39	1997R	0.657	0.466	23.83	5082	27.5	1.1303	22.77	3114	50.7	57.4
40	1997G	0.763	0.513	24.47	6823	22.6	1.1651	23.14	3703	48.4	56.4
41	1996cl	0.828	0.539	24.65	7413	21.8	1.1875	23.16	3728	51.5	61.2
42	1997ap	0.83	0.540	24.32	6368	25.4	1.1882	22.82	3195	60.3	71.6
					Mean	32.2			Mean	47.9	52.0
					+/-	1.00			+/-	1.2	1.23

Chapter One

Deriving the Hubble Equation

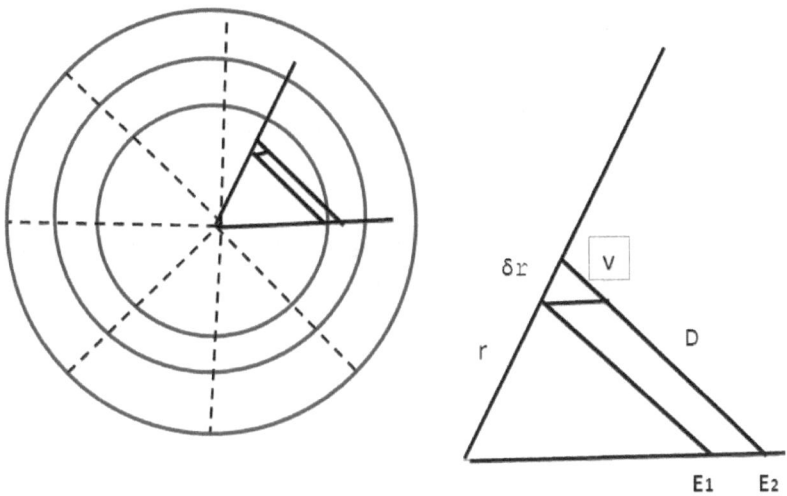

Figure 1.A1.

The figure represents a circle (or a section across the universe) whose radius is expanding regularly with time. The enlarged section shows that light travelling from one radius to another means that the photon travels a distance. A period of time later, the distance has increased as has Earth's position, E1 to E2, on its particular radius. The change in distance for that period time is velocity. If that period of time is 1 second, this is velocity per second. The vertical and horizontal axes of the enlarged section are both distance and time.

The constancy of the ratio between velocity and distance implies that there is a simple mathematical equation that describes this relationship. If the radius of an expanding sphere is R and 1 second later is R + i, then the mean increase in the radius for that second is i/2. Similarly for a fraction of that radius (r in Figure 1.A1), the mean increase is δr/2. Equally the mean velocity of a galaxy for that second is v/2. From Euclidean geometry of similar figures, this results in the following:

Equation Appendix 1.5 $(i/2)/R = \delta r/2/r = (v/2)/D$
Equation Appendix 1.6 $i/2 = Rv/2D$

The volume of the sphere for that second then becomes:

Equation Appendix 1.7 $Vol = 4\pi (R + i/2)^3$

Expanding Equation Appendix 1.7 and ignoring the third and fourth parts of the cubic expansion as being too insignificantly small, as their dividers are D^2 and D^3, where D is in distances of many light years, and v is in fractions of the velocity of light, results in:

Equation Appendix 1.8 $Vol = 4\pi (R + 3Rv/2D)^3$

Therefore the average volume 1 second beforehand becomes:

Equation Appendix 1.9 $Vol = 4\pi R^3(1-t(3v/2D))$
where t equals 1 second.

When the volume is effectively zero, t is the age of the universe at the time the galaxy being observed emitted the light that we now see. This results in:

Equation Appendix 1.10 $0 = 1-t(3v/2D)$

from which:
Equation Appendix 1.11 $Ho = v/D = 2/3t$

But $t = Age - D_t$

where D_t is the time taken for light to travel from the galaxy in question to reach Earth. This results in:

Equation Appendix 1.12 $Ho = v/D = 2/3(Age-D_t)$

D is in light years and v is in fractions of the velocity of light. Converting these into the customary units for the Hubble constant, that is in km/sec and mega parsecs, results in:

Chapter One

Equation Appendix 1.13 $\qquad H_o = v/D = 652/(Age-D_t)$

Since D_T will be different for different galaxies, under Euclidean geometry the Hubble constant cannot be constant. But as shown in Tables 1. and 2, the Hubble constant is constant relative to its time frame!

The only way Ho can be constant is if one side of Equation Appendix 1.9 is multiplied by the ratio of times:

Equation Appendix 1.14 $\qquad v/D = (652/(Age-Dt)) \times (Age-Dt/Age)$

Thus if D (the distance to the observed galaxy) is so far that it takes half the age of the universe for its light to reach us, then the observed velocity is twice what it should be. But the observed velocity depends upon the red shift. It follows that the time frame for that red shift was twice that of our present Earth time frame. Time was twice as fast then. There were 2 seconds for every 1 second now. The pace of time was twice as fast. Likewise if the age (D_T) was a quarter of the age of the universe (in Earth time), the pace of time was 4 times its present value. It follows that the universe has its own system of time, herein after called cosmological time. In cosmological time the period of the second, or the pace of time, is expanding exponentially. Present Earth time relies on the belief that the pace of time has been constant since time began. There is no evidence to support this belief. In contrast, the Hubble value can only be constant if time is expanding. The observational evidence shows that this is the case.

CHAPTER TWO

A Question of Time

The constancy, up to the edge of the visible universe, of the Hubble constant creates significant difficulties concerning time. The Hubble constant enables a calculation of the age of the universe. But this calculation is based on the assumption that time, the period of the second, is and always has been constant since the beginning of time. There is no evidence that this assumption is true, although it is a widely held belief. This belief is based on extrapolating from the fact that since mankind first attempted to measure time, the period of the second, within the limits of measurement, was always the same. There are various means of measuring and so defining the second, from the period taken for a crystal to vibrate a number of times, to the internal motions of an atomic clock. But they all rely on some arbitrary number. The most logical definition of the second, based on one of the fundamental constants of nature, is that period of time for a photon to travel, in space or in a vacuum, 3×10^8 m (actually 299,792,458 m, but within the accuracy of astronomical measurements, 3×10^8 m is good enough).

There are several pieces of evidence which show quite clearly that this assumption must be wrong, as the age of the universe is well beyond that predicted by the Hubble constant (whether the age is 12.5 billion or 13.7 billion years is immaterial).

The first piece of evidence comes from observations made about the globular clusters. These are clusters of stars, each cluster having up to several thousand stars that are, in astronomical terms, very close to each other in age. There are over 150 such clusters within our galaxy. The stars in any one cluster have the same magnitude, and spectral analysis shows that they have the same composition. The composition varies slightly from one cluster to another, but it always shows that the stars in any one cluster are extremely old, with some clusters being more than 14 billion years old. This creates a paradox about time.

Chapter Two

The second piece of evidence comes from the Sloane star survey. This is a survey of the most distant galaxies all around the universe. Although the survey is not quite complete, it shows that our galaxy is surrounded by galaxies on all side that are more than 10 billion light years away. But mathematically the sum of the distances in billions of light years from our galaxy to galaxies that are on opposite sides of the universe cannot exceed the age of the universe in billions of years. If the universe arose from a single location, it would take time for material to get to those distances, form stars and galaxies, and go through the life cycle of big stars to create a supernova (used to detect their distances), and then for the light from those supernovas to travel back to our galaxy. Early analysers of the Big Bang theory proposed that there was a period of inflation whereby proto-matter moved at velocities greater than the speed of light, whilst the universe itself expanded at a speed greater than light. This theory was put forward to explain the large scale uniform structure of the universe. Inflation, it has been calculated, ceased when the universe had a diameter of only a few metres. This theory has a touch of magic about it, where magic is a phenomenon that apparently cannot be explained by the known rules of physics. Obviously, as the rules become more known and science progressed, so things attributed to magic were reduced. This is a continuously on-going process. Thus a seventeenth-century man seeing someone switching on an electric light would consider it magic, as it would be beyond his knowledge of science. Since we cannot be sure that we know all the rules of physics, it is remotely possible that there was a period of inflation or something very similar. However, no mathematical analysis has ever suggested that any episode of inflation expanded the universe to a diameter of billions of light years. That is, the sum of distances to galaxies on opposite sides of the universe creates another paradox.

The third piece of evidence arises from all theories which posit that the universe arose from a single position and has been expanding ever since. When looking through telescopes at galaxies that are billions of light years distance from us, we are looking back in time. The light from a supernova 10 billion light years away that is only just reaching us means that the supernova occurred 10 billion years ago. But then the universe was very much smaller, which means that the galaxy with the supernova would have been much nearer. So why has it taken so long for the light to reach us, especially if the speed of light is constant? This is the distance paradox.

It follows that the universe must use a different system of time, whereby the period of the second has not always been constant. This in turn affects how to interpret the signals within the light from the very distant galaxies.

Light, Wavelength, and Frequency

White light consists of a mixture of colours that make the rainbow colour spectrum. Each colour is carried as photons in a band of particular electromagnetic frequencies. The frequencies merge seamlessly through the colour spectrum and beyond, with slower frequencies being in the infrared zone and higher frequencies being in the ultraviolet range. It takes a very brief period for the emission of a photon, that period depending upon the frequency. Thus the photons for red light have a frequency band width of 4.28 to 4.63×10^{14} hz, whilst photons from the colour violet have a bandwidth of 7.11 to 7.49×10^{14} hz. The inverse of the frequency is the time taken for the emission of the photon. It is of the order of one hundredth of a million millionths of a second.

If the object emitting the photons is moving very fast, then the last part of the photon's wave has to travel a longer distance; that is, the distance the object has moved during that minute fraction of a second. This lengthens the wave and so reduces the frequency. If the velocity of the light emitter is fast enough, the photons at the extreme end of the red light bandwidth will be stretched enough to become infrared waves. Similarly, some waves in the orange colour bandwidth will become stretched to become waves for red light, and so on across the whole spectrum.

At the other end of the spectrum, some of the ultraviolet waves will become stretched to become waves we see as the colour violet. Overall and for velocities reasonably less than relativistic velocities, the balance of energy between the different colours is virtually unchanged, and we see white light. The intensity of any colour is determined by the amplitude of the wave. For comparatively fat waves, the stretching makes little change in the amplitude of the wave. But for thin spiky waves such as ultraviolet waves, the stretching reduces the amplitude far more. Given the width of each of the colours, if there has been recruitment of many ultraviolet waves as a result of extremely high velocity of the light source, the brightness

Chapter Two

contribution by the colour violet will lessen, allowing the photons for red colour to gradually dominate the white light. This is red creep, which is distinct from red shift. Red creep only starts to develop at velocities well above 76 per cent of the velocity of light.

Light from incandescent elements has a much more specific structure. For any particular element, there is a major or dominant set of frequencies of a very narrow bandwidth plus several other sets of narrow bandwidth that are less bright. Between these bands there is very little to no light. A negative photograph of the spectrum shows these striking patterns of bands. Since each element has its own set of bands, it is possible to deduce the composition of the incandescent material. The intensity of light at the major frequencies allows for a reasonable estimate of that element's concentration in that incandescent material. If the material is moving away, the wavelengths lengthen, reducing their frequency so that on photographs of the spectrum, the bands have shifted towards the red end of the spectrum. This is the red shift and allows determination of the velocity of the incandescent moving object.

Of major significance, once the photon has been released, it travels at the speed of light, and its wavelength cannot be altered. The information contained in the red shift cannot be altered. This has important significance when considering the background microwave radiation.

Stretching of electromagnetic waves during transit has been proposed, although there is no objective evidence of this. In any case, stretching forwards in the direction of travel would mean that part of the photon travels at a speed greater than light, which is impossible. Stretching rearwards would displace and slow the succeeding wave. In a long train of waves, this would mean that photons at the tail end of the train would be travelling slower than the speed of light, which again in space is impossible. Alternatively, the tail end of the stretched wave could be absorbed by the succeeding wave, but in that case the frequency and wavelength have not been changed, that is, there has been no external evidence of any stretching. Photons, as waves, do stretch up and down and sideways. The width of the wave expands at the velocity of light, and this demonstrates the inverse square law, whereby brightness is reduced but the light covers a great area.

With a constant wavelength, the frequency is dependent on the speed of light, c. This is a fundamental constant, and by definition fundamental constants are constant in all time frames. If time should quicken, the speed of light (as judged in Earth time) will quicken. Conversely if time should slow, the speed of light will slow. There is another medium whereby the speed of light is slowed. This is when light is passing through water; in water the velocity of light is almost half its velocity in a vacuum. Such a reduction in velocity causes the waves to bunch up. This causes the amplitude of the wave to increase. That is, the light appears bright. Colours generated under water (e.g., coloured tropical fish) appear much brighter whilst the fish are in the water.

In the previous chapter, it was shown that when time is slowed due to the special relativity effect on time, the rate of photon emission is reduced, resulting in a dimmer light and so an increase in magnitude and a false exaggeration of the distance and hence a smaller result in the calculation of the Hubble value. Correction for this results in an improved Hubble value. This is quite distinct from the bunching up effect of slowed transit velocity. Here the speed of light is reduced, but so is time. The rate of supply of photons is reduced, and so there is no pressure to bunch up and enhance the brightness.

Conversely, if the pace of time is faster, there will be an excess of photons arriving at the observer's location. This will result in an increase in the perceived brightness and hence a lower magnitude and an underestimate of the distance to the source of the photons. Correction for the pace of time will result in a correct calculation of the Hubble constant. This is applicable to all the supernovas. It is this that underlines the concept that in the past time was much faster than it is at present.

The equation for calculating the age of the universe in Earth time, given in chapter 1, shows that the correction must be the age of the universe in Earth time when light was emitted from the supernova divided by the age of the universe in Earth time. If a Type 1A supernova erupted when the universe was half its present Earth time age, the calculated distance to that eruption, which is based on brightness, would need to be doubled to get the correct Hubble value.

Chapter Two 23

The brightness is based on the rate of production of photons. There would be twice as many as expected in Earth time. It follows that the pace of time was double that of the present Earth-based pace of time. There were then 2 seconds where now we have only 1.

Similarly, a supernova occurring when the age of the universe, in Earth time, was a quarter of the universe's present Earth time age, the pace of time would be doubled again, and so on as one goes further back into the history of the universe. Since the beginning of time, every time its Earth time age doubled, the pace of time was halved.

Thus a key principle emerges: The Hubble constant can only be constant if time has been continuously expanding or slowing since the beginning of time.

The Hubble or acceleration constant is another fundamental constant of nature and therefore is constant in all time frames. That is, the very fact that it is constant confirms the key principle that there has always been a continuous expansion or slowing of time. The universe, or cosmos, has its own system of time, hereinafter called cosmological time.

The ratio of ages used as the correction factor enables an estimate of the precise relationship between cosmological time and Earth time. If the Earth time age is 12.5 billion years, then over the last 6.25 billion years there has been 8.84 cosmological years. This quantity of time is constant for every period that the Earth time age doubled. That is, cosmological time has been slowing exponentially to converge to our present pace of time.

Nature has many examples of exponential slowing. The half-life of a piece of radioactive material is a prime example. Similarly the rate of cooling of a hot object, or the rate of flow when the tap of a water butt is opened, or the rate of washout of a pollutant in a coastal harbour are examples of exponential change. Each change is dependent on the previous value.

It follows that if, when working backwards, one knew how many occasions the pace of time has doubled since the beginning of time, one could calculate the age of the universe in cosmological time and so the present rate of change of cosmological time. The latter is of supreme importance and is discussed in the next chapter.

CHAPTER THREE

Time, Mass, and the Age of the Universe

To accelerate a star such as the sun, mass ~2 × 10^{30} kg, requires a lot of energy. But this is insignificant compared with the energy required to accelerate a black hole. But that too is insignificant compared with the energy needed to accelerate a whole galaxy, and there are many billions of galaxies being accelerated continuously. The amount of energy needed for this is prodigious. There is only one mechanism known to produce energy on this scale. This is from the atom. From Einstein's famous equation e = mc^2, 1 kg of mass produces a substantial amount of energy. But the energy requirement to accelerate all the galaxies, and accelerate them continuously, is of the order of the energy equivalent of the mass of whole stars per second. We are only familiar with mass-to-energy conversion occurring as violently explosive events, yet the acceleration of stars and galaxies is a smooth continuous process that goes on and on, unchanging, over billions of years. There must be a mechanism whereby mass is converted to energy smoothly and continuously and non-explosively and is not confined to specific locations. That is, the energy must come from the individual atoms, and the energy released must be in a controlled smooth fashion at a more or less uniform rate. There are exceptions, such as the violent release of energy from supernova explosions. But these are rare events, occurring about once in every 200 to 300 years per galaxy.

The mathematical imperative that the Hubble constant is constant to the edge of the visible universe (and it is) means that time has been slowing continuously and smoothly since time started. This points to a solution. As time expands (slows) some of the mass of the atom is released as energy, just as the atom releases gravitational energy. In a large mass of atoms the overall mass must be reduced. That is, mathematically, time and mass are inversely related.

Chapter Three 25

Curiously there are a number of equations in physics (general physics, as well as the physics of the very small, quantum physics, and the physics of the very large, the physics of relativity) which all show this inverse relationship.

Since Galileo several hundred years ago first discovered the properties of the swinging pendulum, until comparatively recently, the pendulum has been the mainstay of measuring time. Only in modern times has it been overtaken by the atomic clock and the number of vibrations of a crystal. The key to its long popularity is that fact that the period of the pendulum depends upon two factors: the length of the pendulum and the gravitational force. The period is proportional to the square root of the length of the pendulum divided by the gravitational force. It is the Earth's gravitational force which keeps the bob of the pendulum swinging.

Gravitational force is the force that exists between two masses that are a distance apart. The numerical value of the force is the mathematical product of the two masses multiplied by the gravitational constant (another fundamental constant that is constant in all time frames) and divided by the square of the distance between them. The square root of two masses multiplying each other is mass. Going back to the pendulum equation, the period of the pendulum swing is thus proportional to the inverse of mass multiplied by the residua of the rest of the pendulum equation. But that period quantifies time. If time should expand, and the laws of physics maintained then mass must be reduced, and reduced by the same proportion as the increase in (or expansion of) time.

A similar situation exists in quantum physics. There is a very simple equation which says that the energy of an electromagnetic wave is proportional to the frequency of that wave. But energy equals mass multiplied by the square of the velocity of light, the famous $e = mc^2$. Frequency is a dimensionless number divided by time. That is, mass is proportional to the inverse of time. This is merely a reciprocal of the time mass relationship shown in the pendulum equation. If time increases, mass must be reduced.

Relativity theory rests on a foundation that the laws of physics are the same for all observers, wherever they may be. But it was from this principle that Einstein was able to derive the special relativity equation, which showed that time slows with increasing velocity, and from this he derived his famous $e = mc^2$ equation. This leads to the famous twin paradox. This goes that if the elder of twins made a round trip to Alpha Centauri—a star 4 light years away—travelling at an average speed of half the speed of light, when he returned, his stay-at-home younger sibling would be 18 years older but the traveller would, according to his own clocks and calendar, be only 15.6 years older, that is, he would now be almost 2.5 years younger than his younger twin. The calculations for this follow a precise protocol. Einstein originally thought that exactly the same mathematics applied to fast moving masses. This contradicts the very foundation of relativity theory—that the laws of physics (and so their associated equations) are constant.

This may be shown by using one of Einstein's thought experiments. How fast would a pendulum swing if it was in a long case clock in a cave on the moon if the moon then started orbiting the Earth at relativistic speeds? According to the special theory of relativity, time would slow and the pendulum would therefore slow. But if mass increased with velocity, the gravitational force generated would be increased, and this would shorten the period of the pendulum. Which predominates? That is, the law of physics as it relates to pendulum theory would be broken. But fundamental to all relativity theory is that the laws of physics are the same for all observers. The only way to resolve this contradiction would be if mass, and its gravitational strength, decreased by exactly the same proportion as time expanded. That is, time and mass are inversely related.

There are three interesting consequences of this. If the velocity becomes equal to the velocity of light, the second is expanded to become infinitely long. As a consequence mass becomes zero. If mass becomes zero, it cannot be accelerated. Electromagnetic waves are energy which is a form of mass; that is, electromagnetic waves, light, cannot be accelerated further. The speed of light is therefore the maximum possible speed. Nothing can go faster than the speed of light. Speed of course is relative to the pace of time. If time is faster, the second shorter when compared with our time, the speed of light will be faster. This is repeating that the speed of light is constant in all time frames.

The second consequence is that it defines the lifetime of a star or a galaxy. The Hubble constant is 5×10^{-10} m/sec/sec. It should take exactly 19 billion years for a star or galaxy to be accelerated to the speed of light, at which point its mass will have disappeared. But it is not as straightforward as that. By the time the acceleration of that galaxy had reached half the speed of light, relativity time dilatation would interfere. Time for the galaxy would have been stretched to 1.15 times its ambient value. This would reduce the acceleration force from 5×10^{-10} m/sec/sec to 4.33×10^{-10} m/sec/sec. At a velocity of three quarters of the speed of light, the acceleration would be 3.3×10^{-10} m/sec/sec. At a velocity of 0.99 that of light, the acceleration would be 0.7×10^{-10} m/sec/sec. To a hypothetical passenger in that fast receding galaxy, it is still being accelerated at 5×10^{-10} m/sec/sec. To a distant observer the galaxy's acceleration is slowing, so prolonging the life of the galaxy. Eventually the galaxy would almost reach the speed of light, but the acceleration required to reach that speed would be so low that it take an enormity of time but still never quite reach light speed (but its mass would have been reduced to fine dust). The light output from the galaxy would have long diminished to become undetectable. To us on Earth, we would see its light dwindling, and perhaps more importantly, the red shift, which would have been increasing with increasing velocity, would slow down its rate of displacement in the light spectrum in that light. This would appear to have been observed and led to the conclusion that the expansion of the universe is slowing down. The ultimate conclusion is that the Hubble constant is indeed constant in all time frames, but because of time slowing at relativistic speeds the life of a galaxy is longer than 19 billion years of cosmological time, and galaxies are fated to become faint dust, forever receding from the visible universe at a velocity fractionally less than the speed of light.

The third consequence is that neutrinos, which travel at the speed of light, must be massless. There has been dispute about this, but the very velocity of neutrinos rules out any mass.

The Age of the Universe in Cosmological Time

It has been calculated that at the beginning of the universe, 10^{80} nucleons were formed. From Avogadro's hypothesis of the number of molecules in a mole of hydrogen, 1 g molecular weight of hydrogen (2.016 g) contains 6.02×10^{23} nucleons. The sun, mass 1.98×10^{30} kg, contains 5.8×10^{56} nucleons. It follows that the original mass of the universe was 3.39×10^{53} kg. A black hole has an event horizon where the gravitational force is such that to overcome it requires a velocity greater than the velocity of light. The radius of the event horizon is proportional to the square root of its mass. If the Earth were to become a black hole, it would have an event horizon radius of 1.15 km; for the sun this would be approximately 660 km. The original mass of the universe, which was concentrated into one location, would have formed a primeval black hole. For a mass of 3.39×10^{53} kg, the event horizon of this primeval black hole would be 28.8 light years. This assumes that the pace of time was that of the Earth's present pace of time.

But the pace of time in that early era was much faster. From the inverse relationship between time and mass, the mass would have been much greater and the event horizon radius much longer. It would have been $2^{N/2}$ longer, where N is the number of doubling periods. It would have taken light $28.8 \times 2^{N/2}$ years to get from the centre of this primeval black hole to the event horizon (see Figures 3.A1 and 3.A2). This hypothetical time then suffered 2^N doublings to equal the 12.5 billion years, deduced from the Hubble constant as the age of the universe in Earth time. It is then simple arithmetic to calculate the value of N as being 19.1. Each doubling period is 8.84 billion years of cosmological time. That is, the age of the universe in cosmological time is 169 billion years or approximately 5×10^{18} cosmological seconds. This figure assumes great significance when calculating the energy output of the sun or the gravitational energy output of the moon (see later chapters). The primeval black hole would have had a radius of some 23,000 light years. This may be compared with the radius of our galaxy, which is ~50,000 light years.

The 19.1 doubling figure means that when time started, the pace of time was approximately 650,000 times faster than it is now. The period of a second was 1/650,000th of its present duration. A year of our Earth time would be approximately 6 sec of cosmological time then. Light would have travelled 650,000 times faster than it does now, although relative to the pace of time then, it was travelling at 3×10^8 m/sec.

If the cosmological age of the universe is 169 billion years and the lifetime of a recognisable galaxy is around 20 to 25 billion years, there must have been a considerable turnover of galaxies, as those galaxies were accelerated away to almost nothing by the Hubble acceleration force. This in turn means that the evolution of the universe has been a very long and slow affair. From that primeval black hole, galaxies have come and gone to be replaced by more emerging mass (that in turn formed galaxies) from that primeval black hole. This renders the Big Bang theory obsolete. The current generation of galaxies we see in the visible universe must be less than 20 to 25 billion years old, but the universe is very much older than that.

Two profound philosophical problems remain. One is the spread of time. Time originally was very fast around the slowly collapsing primeval black hole. A long way distant from that primeval black hole, time did not exist. This suggests that time spreads like light into empty space. The alternative view is that space also was created then, forming a space-time continuum (whatever that phrase might mean). The problem with the latter is that it denies a logical cause for the start of the collapse of that primeval black hole—which is the loss of gravitons radiating out into empty space such that the event horizon had to recede into the black hole, allowing the gradual escape of proto-mass and other forms of energy. If the rate of spread of time is not infinite, then events separated by a distance cannot be simultaneous—a problem often discussed in books on relativity but outside the scope of this book.

The other problem is that if time and mass are inversely related, and mass is merely a compacted form of energy, then time is 1/energy. If energy is the ability to do work, its reciprocal has no meaning in terms of physics.

APPENDIX 3

A. The Time Mass Relationship

Time and mass are inversely related. This can be seen in three different realms of physics.

(i)   The pendulum equation has the following form:

Equation Appendix 3.1   time = $2\pi\sqrt{(l/g)}$

where time is the period of the pendulum, and l is the length of the pendulum and g is the gravitational force. But:

Equation Appendix 3.2 $g = GM_1M_2/r^2$

where G is the gravitational constant and is constant in all time frames, and M_1 and M_2 are two masses (kg) separated by distance r (metres). It follows that:

Equation Appendix 3.3   time = $(2\pi r/M) \times \sqrt{l/G}$

That is, time and mass are inversely related.

(ii.) In quantum physics, the energy of an electromagnetic wave is given by:

Equation Appendix 3.4 $E = h\lambda$

where λ is the frequency of the wave and h is Planck's constant that is also constant in all time frames. It follows:

Equation Appendix 3.5 $mc^2 = h \times n/time$

That is, time is inversely related to mass.

(iii.) In relativity physics the two relevant equations are:

Equation Appendix 3.6 $T = t \, (1/(1-v^2))^{0.5}$ and

Equation Appendix 3.7 $m = M(1-v^2)^{0.5}$

where v is the velocity expressed as a fraction of the velocity of light. For Equation Appendix 3.6 and Equation Appendix 3.7 the uppercase T and M are larger than t and m, respectively. It follows that:

Equation Appendix 3.8 $(1-v^2)^{0.5} = t/T = m/M$

Equation Appendix 3.9 $tM/T = m$

That is, if time T expands (that is, slows), the mass m gets smaller.

B. The Doubling Time

The halving of the pace of time with each doubling of the Earth time age of the universe describes two exponential series of numbers. This enables the two to be depicted graphically as a log-log graph (Figure 2). The triangles marked with an X show that the increase in cosmological time is constant for all doublings of the age of the universe. Thus from the graph the average number of seconds corresponding to one Earth second over the period when the Earth time age increased from 6.25 to 12.5 billion years has a log value of 0.15 or 1.41. That is, the cosmological age increased by 6.25 billion × 1.41, or 8.84 billion years. This same number applies to the occasions when the Earth age doubled. This is the doubling time. The age of the universe in cosmological time, then, is this number times the number of occasions that the Earth time age doubled.

C. The Age of the Universe in Earth Time

Equation 1.1 in chapter 1 predicts that the age of the universe is 652/Hubble constant in billions of years. That is, the age of the universe should

Chapter Three 32

be 12.5 billion years in Earth time. In practice it must be older, as near the limits of the expansion, the velocities of the receding galaxies are so high that the galaxies start to suffer from relativistic time slowing.

In theory, those fast moving galaxies never reach the velocity of light, their time slows so much, although any traveller on such a galaxy would still think he is moving with an acceleration of 5×10^{-10} m/sec. But long before then, the mass of the galaxy would have diminished to dust. Even before then, the light from the galaxy would have been virtually extinguished; that is, the galaxy would no longer be part of the visible universe. It is possible that the time slowing would have added up to a billion years to the age of the visible universe, but this is a wild estimate. The age of the visible universe is therefore between 12.5 and 13.5 billion years of Earth time.

D. The Age of the Universe in Cosmological Time

The universe has an estimated 10^{80} nucleons. From Avogadro's hypothesis, this would give the mass of the universe as 1.67×10^{53} kg. If all this mass was concentrated in one place, it would make a primeval black hole. Black holes have an event horizon (or Schwartzchild radius) where the strength of the gravitational field prevents light and all radiation escaping. The formula describing this radius is given by the equation:

Equation Appendix 3.10 $r = 2GM/c^2$

where G is the gravitational constant, M is the mass of the universe in kilograms, and c the velocity of light in metres per second; r, then, is the radius of the event horizon in metres. There are serious problems with this equation when applied to the beginning of the universe, when all mass was concentrated in one location. The radius becomes 2.48×10^{26} metres, or 26 billion light years, more or less equal to twice the radius of the observable universe. The mean density would be approximately one proton per cubic metre. But by definition, a black hole is made of material of an extremely high density.

Equation Appendix 3.11 $g/kg = GM/r^2$

Chapter Three   33

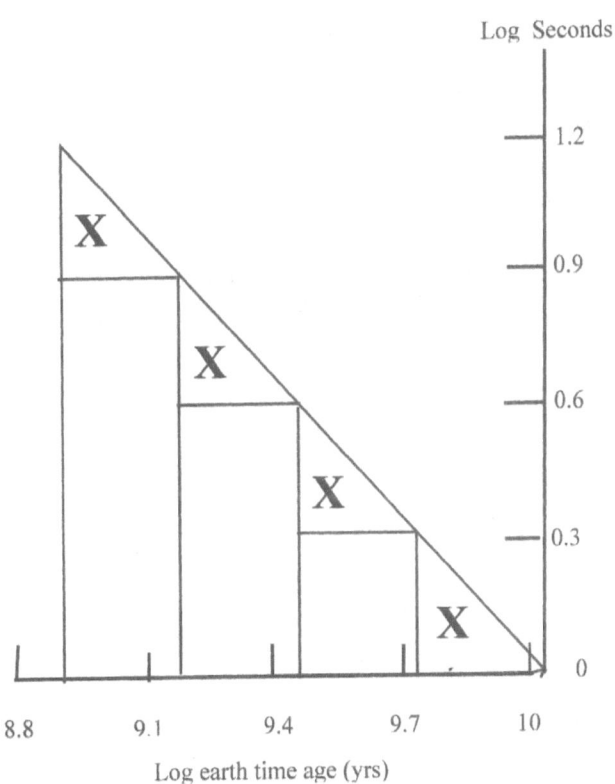

Figure 3.A1. The relationship between Earth time and cosmological time. The vertical axis is the logarithm of the number of seconds of cosmological time per Earth time second. The marked triangles are equal areas and are a reflection of the fact that cosmologically, the doubling time is constant for each occasion that Earth time (age) doubled.

Equation Appendix 3.11 describes the acceleration due to gravity. Acceleration can never exceed c per second, as whatever was being accelerated would be travelling faster than the speed of light within a second. Gravitational acceleration is essentially negative; that is, it accelerates towards the source of the gravitational energy. Photons travel at a velocity of c away from the source of light emission (and gravity). It follows that at the event horizon, the two velocities cancel each other out. From Equation Appendix 3.11, the event horizon can be calculated from:

Equation Appendix 3.12 $\quad r = (GM/c)^{0.5}$

where G is the gravitational constant, M is the mass of the universe, and c is the velocity of light; r, then, is the radius of the black hole. Assuming no change in the pace of time it would have taken 28.8 years for light to travel from the centre of this primeval black hole to reach the event horizon. This centre can be equated with the singularity that was the starting point of the expansion of the universe, as postulated under the Big Bang theory. Since then the radius has doubled repeatedly to become 12.5 billion light years. Time, that is age, must also have doubled repeatedly.

But at the beginning, time was much faster, which would have increased the mass and so enlarged the event horizon radius. The mass would have doubled for each doubling of the Earth time age. For N doublings, mass would have increased by a factor of 2^N. The resulting change in the diameter of the primeval black hole would then be:

Equation Appendix 3.13 $\quad r = 28.8 \times 2^{N/2}$

where r is in light years. Since then it has doubled itself N times. Cosmological time also started then, and its pace was very fast. It has been expanding (slowing) exponentially ever since (see Figure 3.A2 see below) to become almost equal to the pace of Earth time. This occurred at the Earth time age of the universe, that is:

Equation Appendix 3.14 $\quad 28.8 \times 2^{N/2} \times 2^N = 12.5 \times 10^9$ years

This resolves to:

Chapter Three 35

Equation Appendix 3.15 $\quad 3N/2 = (\log(12.5 \times 10^9) - \log 28.8)/\log 2$

That is, there have been 19.1 (N) doublings since time first started. But each doubling is worth 8.84 billion years of cosmological time, and this yields the age of the universe in cosmological time as 169 billion years, or 5.33×10^{18} cosmological seconds. (If one takes the Earth time age of the universe as 13.5 billion years, there have been 19.3 doublings, giving the cosmological age as 169.75 billion years or 5.35×10^{18} sec.)

The radius of the primeval black hole would then have been $28.8 \times 2^{N/2}$ or 22,350 light years. Applying the time correction to the Schwartzchild equation (Equation Appendix3.10), the equation apparently becomes:

Equation Appendix 3.16 $\quad R = 2GM \times 2^N /(c \times 2^N)^2$

$$= 42,313 \text{ light years}$$

But this figure assumes that G, the gravitational constant, is unaffected by expanding time. But with time expansion, the Newton must decrease by the square of the change of time. That is, at the beginning G had a value of $6.67 \times 10^{-11} \times (2^N)^2$. This makes, according to the Schwartzchild equation, the event horizon $42,313 \times 2^{38.2}$ light years—an impossible number. It follows that the Schwartzchild equation cannot be applied to events occurring at the origin of the universe.

Chapter Three 36

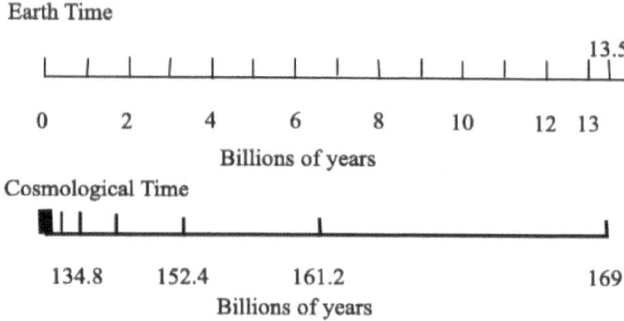

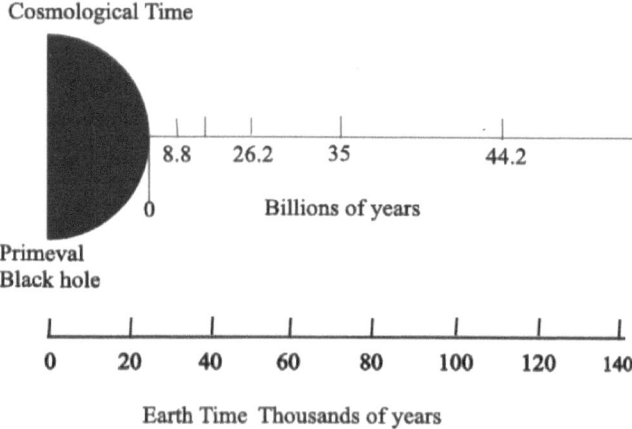

Figure Appendix 3.2 A comparison of Earth time with cosmological time. The lower section is a comparison over the first few thousand years of Earth time.

CHAPTER FOUR

The Sun

The stars in our galaxy were originally classified according to their spectra using an alphabetic system, A to Q (omitting J). This has been simplified to seven letters, O, B, A, F, G, K, M, with numeric subdivisions. G stars are among the most common stars in our galaxy, and our sun is a G2 star.

The sun has a mass of slightly less than 2×10^{30} kg, or 2 times a thousand billion billion billion kilograms. It is half a million times more massive than the Earth and approximately a thousand times more massive than Jupiter. The sun's surface temperature is around 6,000°C, although its internal temperature is estimated to be over 10 million degrees. The sun's radius is 2.32 light seconds (almost 700,000 kilometres). Its average relative density is 1.4, compared with water, which has a value of 1.

Gravitational compression is responsible for the density, but clearly at the centre of the sun the density must be at least 100 times this. This aspect assumes major significance which is discussed later. Ordinarily a density greater than 1 implies that the substance is nongaseous, but at the centre of the sun it is so hot that everything is vaporised. The sun is composed of 75 per cent hydrogen and 24 per cent helium, the remainder being a mix of a number of elements. The sun's estimated age is around 6 billion years: approximately half the age of the universe when that is reckoned in Earth time.

The sun was created from the remnants of a giant supernova which spawned not only the sun but all the material that makes the solar system. That supernova was the consequence of a giant star, 50+ solar masses, which had used up the majority of its hydrogen that was in the centre. As a result the central temperature rose, triggering helium fission. The outer coat still contained a high concentration of hydrogen but was heavily contaminated with helium as a consequence of convection currents. The resulting helium fission explosion blew part of the outer coat as several blocks. Gravity then caused the largest block to condense into a sphere, the proto-sun. The supernova also created all the other elements in the periodic table, some of which travelled with the proto-sun. Gravity caused the heavier atomic elements to be drawn near the proto-sun, with much of that material falling into the sun. If 1 per cent of the solar mass is composed of these elements, this has over 100 times more of these heavy elements than the combined masses of all the inner rocky planets.

How big was the sun's helium dowry? This is impossible to say with any certainty, although it must have been substantial. A reasonable estimate is that at least 10 per cent of the sun's mass consisted of that helium dowry, if only because Jupiter, which originated from the same source at the same time, has 10 per cent helium but has no means of synthesising it. This means that up to 14 per cent of the sun's solar mass, or 3×10^{29} kg, of helium has been synthesised from hydrogen in the sun's 6 billion Earth time years. If the sun's age in cosmological time is 8 billion years it is currently producing 1.2×10^{12} kg, or over a thousand billion kilograms, of helium per second.

The sun's classification as a G2 star is based on its temperature and spectral composition, which reflects the mean concentration of helium. The inference is that all G2 stars have a significant helium concentration well over and above what could be expected from any exothermic hydrogen fusion. This implies that all G2 stars were formed from previous supernovas of giant stars and in the process were given a handsome dowry of helium. G2 stars are very common in our galaxy. This is so unlikely as to question the very validity of exothermic hydrogen fusion. This then questions the mechanism of the sun's thermal output. But there are additional reasons for questioning the hydrogen fusion concept as the source of the sun's thermal output.

The mechanism behind the sun's thermal output has been a matter of speculation for centuries. In the nineteenth century it was realised that there was no chemical process that could produce so much energy for more than a few thousand years. Einstein's discovery of special relativity, and particularly his discovery of his famous mass energy equation, showed that the destruction of mass produces vast quantities of energy. This seemed to be the answer. This was particularly so when it was realised that four atoms of hydrogen had a greater mass than one atom of helium. This led to the natural conclusion that it was hydrogen fusion that lay behind the sun's thermal output. There was no attempt to see if there was any other evidence that justified this conclusion. In fact the evidence points in the other direction.

There is an oddity about the sun and in fact about all the stars. This is their remarkable stability; at least until their hydrogen stores are seriously depleted. The geological record of the Earth shows that for at least 1 billion years the sun's thermal output has been stable to within a few degrees, apart from occasional blips of the ice ages. In that time the sun has produced many billions of tons of helium. Hydrogen fusion is supposed to be a violent event, producing vast quantities of energy. If this is so there must be some controlling factor which ensures that the fusion occurs throughout a body with a diameter of 4.64 light seconds at an extremely controlled rate both in time and in the spatial distribution within that body. This is so outside the realm of physics as to be highly unlikely.

Another reason for questioning the validity of the exothermic hydrogen fusion hypothesis lies in the problem of heat transfer. The Earth-solar constant is 1400 joules per square metre per sec This is when the sun is directly overhead. The mean sun-Earth distance is 149.6 million kilometres. It would take a photon, travelling at the speed of light, close to 500 seconds to travel from the sun to Earth. A hypothetical sphere of radius 500 light secs would have a surface area such that if each square metre was receiving 1400 joules of heat energy per second the solar energy output would be 4.396×10^9 kge/sec (1 kge is 9×10^{16} joules, from $e = mc^2$).

The loss of mass as energy in producing 4.0026 kg of helium (helium's atomic weight) is 0.0296 kg. To produce 4.396×10^9 kge of energy means the production, via hydrogen fusion, of 6.03×10^{11} kg of helium per second. Yet the amount of helium produced, even after allowing for a 10 per cent dowry and the cosmological age of approximately 8 billion years, is 12×10^{11} kg of helium per second. That is, solar heat radiation accounts for only half of the amount of helium actually being produced, if this was from exothermic hydrogen fusion.

On average 1.26×10^{17} kg of the solar mass releases 1 kg worth of energy per hour. At the centre of the sun, where the density is at least 100 times the average, this means that within a sphere of radius 6 km 9×10^{16} joules are being produced every hour. All that energy must be dispersed within the hour. But the neighbouring zones have a very similar density and so are producing energy at very nearly the same rate. The thermal gradient is 0.015°C per metre per second. That is, there can be very little transfer of heat energy from that central sphere to its immediate neighbouring zones. But in another hour an additional 1 kge of energy will be produced. The heat at the central core would rise until all the mass in that central core would undergo a fusion chain reaction. The sun would explode as a supernova. Since this obviously does not happen, there must be some mechanism which prevents overheating; that is, hydrogen fusion cannot be exothermic.

Another piece of evidence that undermines the exothermic hydrogen fusion hypothesis is the energy involved in getting hydrogen atoms to fuse. In the process each of the contributing hydrogen atoms, or rather hydrogen nuclei or protons, loses 0.00735 of its mass. From Einstein's mass energy equation, this fraction of a kilogram represents 6.6×10^{14} joules. Einstein's special relativity theory also shows that mass is reduced at high velocities. Thus it is possible to calculate the velocity, as a fraction of the velocity of light that would cause this reduction in mass. Converting this into metres per second shows that the velocity is a fraction over one tenth of the speed of light, 3.61×10^7 m/sec. The energy required to accelerate 1.008 kg of hydrogen to this velocity is 6.6×10^{14} joules.

Gas molecules have mass and move with temperature. For air it is approximately 0.6 m/sec per degree. For the same amount of energy the hydrogen atoms would achieve a velocity of 3.286 m/sec per degree. Heating hydrogen to 11.2 million degrees produces a velocity of 36.1 million m/sec. This corresponds to a velocity that is a little over 10 per cent of the velocity of light. This corresponds to the estimated internal temperature of the sun. The energy required for 1 kg to be accelerated to this velocity is $\sim 7.5 \times 10^{14}$ joules. That is, the energy required for hydrogen to achieve fusion velocity is slightly over the energy output of the fusion reaction. There is no surplus energy for radiation. But the resulting helium atoms require additional energy in part to convert two protons to two neutrons (neutrons have fractionally more mass than protons), in part to bind the nucleons together despite the internal charge repulsion and in part to be strong enough not to undergo fission when in collision with another hot (11 million degrees hot) and fast moving helium atom. It is only when

Chapter Four 41

the helium atoms are even hotter and moving faster that they have enough momentum to break through the binding energy and either fuse to form other elements or, and more commonly, undergo fission, releasing a huge amount of energy. That is, hydrogen fusion is endothermic.

The evidence against hydrogen fusion being exothermic can be summarised as:

(a) The very long term stability of the sun and of all G stars.

(b) The absence of any controlling mechanism which maintains this stability.

(c) The problem of heat transfer from the centre of the sun and the low thermal gradient at the centre of the sun, where fusion is still apparently continuing.

(d) The energy required to get hydrogen atoms to fusion velocity equals the energy equivalent of the mass loss when hydrogen fuses to form helium.

(e) The rate of helium formation that would result is far too low to account for the sun's helium content.

As such hydrogen fusion cannot be the source of the sun's thermal output. Rather it plays an important cooling effect. If for whatever reason any zone of the sun gets overheated, more hydrogen molecules are accelerated to fusion velocity, and the result is cooling. The end consequence is that the sun's thermal output is extremely stable.

It is small wonder that scientists from around the world, despite over fifty years of trying, have failed to get excess energy from controlled hydrogen fusion. They never will. The success of the H bomb was a lucky chance. The temperature produced by its nuclear triggers not only caused hydrogen to fuse but was high enough to cause helium fission.

The Fusion Fission Principle

The elements thorium, neptunium, and uranium all undergo spontaneous partial fission in a series of steps. Each step releases a helium nucleus, an electron, and a significant amount of energy. These radioactive elements pass through eleven such steps before settling into a stable element: lead or bismuth. It follows that in their synthesis, by fusing with a helium nucleus repeatedly, each fusion step absorbed a significant amount of energy. In addition the noted astronomer Fred Hoyle calculated that all elements in the periodic table that had an atomic number greater than 25 required additional energy in their synthesis. That is, the nuclear fusion which produced these elements was endothermic. More than 75 per cent of all the elements in the periodic table have an atomic number greater than 25. At the time of those calculations, there was no awareness of the need of binding energy; the two intra-atomic forces, the weak and the strong, were thought to be enough to hold the nucleons together. The identification that a helium atom needs a binding force to confine its nucleons to be within the smallest possible space suggests that this applies to all atomic nuclei. If this is so, then a general principle emerges. This is that atomic fusion is endothermic and atomic fission is exothermic.

There may be possible exceptions. This is when the lighter fusing nuclei have between them a greater atomic number than the atomic number of the final product. An example is if the isotopes of hydrogen, deuterium (atomic number 2), and tritium (atomic number 3) can be made to fuse to form helium (atomic number 4), this requires the ejection of a neutron. But free neutrons have a very short life, less than half an hour, before they disintegrate, releasing energy.

Solar Energy

The sheer amount of solar energy means that it must come from some form of mass-to-energy conversion. Atomic fusion has been ruled out, leaving the effect of the expansion of time, through its effect on mass, as the only source of solar energy. The rate of expansion of time means that amount of mass is being reduced to energy is the mass of the sun divided by the age of the universe in cosmological time (see chapter 1) when the latter is expressed in seconds. If the age of the universe is 169 billion years (in cosmological time), the rate is $1/5.33 \times 10^{-18}$ per second. Applying this to the sun (mass approximately 2×10^{30} kg) means that the sun is losing ~3×10^{11} kg of mass per second. That is, the energy produced from the sun by the expansion of time is 3×10^{11} kge/sec, where 1 kge is 9×10^{16} joules. (But radiant energy accounts for only 1.17 per cent of this.)

All this energy is shared between

(1) The energy required to move the sun as part of the expansion of the universe, and a similar amount for the sun's orbit around our galaxy. The latter is the same as that calculated for the outer stars of other galaxies, which was attributed to the gravitational effects of so-called dark matter. The Hubble acceleration constant is 5×10^{-10} m/sec/sec. The Hubble expansion of the universe when applied to the sun's mass requires 25×10^{10} joules/sec, and a similar amount is required as part of the sun's orbital velocity. This is clearly insignificant compared with the 3×10^{11} kge of energy generated by time expansion.

(2) The energy required for the huge convection currents within the sun to lift great masses of helium from the interior towards the surface against the sun's gravity. The amount of energy used for this, although substantial in its own right, is likely also to be insignificant against the 10^{11} kge.

(3) The sun's radiant energy. This amounts to approximately 4.4×10^9 kge, or a little over 1 per cent of the energy released by time expansion. The mechanical energy that is required for hydrogen to reach fusion velocity and form helium is substantial. This mechanical energy is ultimately compensated for by the reduction of the mass of the hydrogen atoms in forming helium. There is no surplus energy from the fusion.

(4) Solar gravity accounts for approximately 97 per cent of the energy produced by time slowing. Details from the calculation of the work done by the moon (see later chapter) in creating the height of the tides in mid-ocean of the Earth support this figure.

(5) Helium formation is endothermic and pays a vital role in maintaining the stability of the sun's temperature. The excess helium in the sun can only have come from hydrogen fusion, with energy being absorbed by the helium nuclei as binding energy. Simple calculation shows that if hydrogen fusion was exothermic, 50 per cent of the energy from hydrogen fusion would be utilised in this way. But there is none available. The energy involved in helium formation is twice the amount of solar radiant energy. The primary source of the energy must therefore come from the effect of time expansion on mass.

The binding energy must be overcome if atomic fission is to occur. This in turn allows an estimate of the temperature that must be exceeded for helium fission to take place. (Helium fission was responsible for the energy released in the hydrogen bomb tests.) This enables calculation of the collision velocity. Since helium nuclei move at a rate of 2.68 m/sec per degree, the minimum fusion temperature can be determined. It is 17.5 million degrees. If the binding energy was only twice that of the radiant energy, the fusion temperature would be approximately 12 million degrees.

Quantifying Gravitational Energy

Gravitational energy accounts for 97 per cent of the energy released from the sun by the expansion of time. This equates to 3.71×10^{11} kge/sec for a mass of 1.98×10^{30} kg. This in turn equates to 0.0164 joules/sec/kg of mass. This must apply to all masses. A billion solar mass black hole would lose, as gravitational energy, the mass equivalent to the mass of our sun every 175 years.

The Fate of the Sun

Our galaxy is surrounded by galaxies that are more than 10 billion light years distant. That is, as part of an expanding universe, our galaxy must travel that distance, which takes time. At a constant rate of acceleration, 5×10^{-10} m/sec/sec, the Hubble constant effect, it would take just over 19 billion years to reach the speed of light, at which point all mass would have been converted to energy. There is, though, an added complication. As the velocity approaches the speed of light, relativity-induced time expansion, or time slowing, develops. This would have the effect of slowing the acceleration so that it would take longer to reach light speed. This would extend the galaxy's life by a few billion years.

The effect of the relativity-induced time slowing would be much greater than the universal time slowing that applies to the entire visible universe. The rate of loss of mass would be much greater, releasing considerable amounts of energy. Most of that energy would be gravitational, but there would be an increase in the radiant energy production. As far as the sun is concerned its hydrogen concentration would still be in excess of 50 per cent of the sun's mass. That is, the endothermic consequence of hydrogen fusion would ensure that the temperature remained unchanged until the velocity was well above 95 per cent of the speed of light.

At a velocity of 95 per cent of the speed of light, time will have expanded by a factor of just over 3. The mass of the sun would have been reduced to one third of its present value. But its gravitational energy output would be reduced to one third of its present value. The solar system would, geometrically, remain the same, as all the mass of the solar system would have reduced to one third of its present value, although relative to our time, orbital velocities would be one third of their present velocities.

At a velocity of 99 per cent of the speed of light, the sun's mass would be reduced to 1/7th of its present mass. The sun's hydrogen reserves would be seriously depleted. The ability to maintain the same temperature would be impaired, and the temperature would start to rise.

At 99.9 per cent of the speed of light, time would have expanded by a factor of 22, reducing the sun's mass to approximately 5 per cent of its present value. The rate of change of mass to energy would be accelerating, and with it, the temperature would be rising rapidly. But before that temperature reached helium fission temperature, the velocity would be very close to the speed of light. The sun's mass would be so small, its volume would have increased substantially due to the high temperature. The enlarged surface area would radiate much more energy, slowing down the approach to helium fission temperature. Eventually the sun would slowly but gently fizzle out of existence in between 15 and 20 billion years.

APPENDIX 4

Table 4.A1 Basic Data of the Sun

Mass	1.98×10^{30}	kg
Sun's diameter	1,392,000	km
Nominal surface	6.06×10^{18}	sq. m
Sun's surface gravitational force	274	m/s/s
Solar-Earth radiation constant	1,400	j/s/m²
Mean distance from Earth	498	light seconds
Shell area radius Earth-sun distance	$4 \times pi \times (498 \times c)^2$	sq. m
Total heat radiation	4.37×10^9	kge/sec
Rate of time change	1.876×10^{-19}	sec/sec
Rate loss of mass	3.73×10^{11}	kg/sec
Radiation as % of loss	1.17	%
Solar helium content	24%	
Helium dowry	10%	
Mass lost per 1,000 kg helium	7.4	kg
Mass lost from radiation	4.37×10^9	kg/sec
Helium formed per second	5.91×10^{11}	kg/sec
Helium formed in 6 billion years	1.09×10^{29}	kg
Expected helium formation	2.785×10^{29}	kg
Deficit	1.69×10^{29}	kg helium
Deficit as % of expected	60.8	%
Mass lost from all helium formation	1.09×10^{10}	kg
All helium formation as a fraction of all loss	2.93	%

Chapter Four 48

Notes and Conclusions

1. The rate of time change is 1/the age of the universe in cosmological time, 169 billion years, calculated in cosmological seconds. This figure is robust in that an error of 2 billion years makes negligible difference in the calculation of that 97 per cent of the energy produced by time slowing is gravitational energy. This figure is confirmed in the very accurate prediction of the height of the high tide in mid-Pacific Ocean islands, described in chapter 6.

2. The 10 per cent of the sun's mass being the helium dowry, that is, the amount of helium passed to the sun at its formation. This percentage is the same as that found in Jupiter, which has no source of helium generation. If all the helium found in the sun was the result of solar exothermic hydrogen fusion, the radiant energy output would be such that Earth would be as hot as Mercury. Alternatively if the initial helium contribution was 19 per cent of the sun's mass then the remaining helium could be the result of exothermic hydrogen fusion, but this does not explain why the sun (and all G stars) has been so stable for so long without any controlling mechanism preventing a chain reaction and stellar explosion.

3. The shell area is the surface area of a sphere with a radius equalling the sun-Earth distance.

4. The total radiation output is calculated in kge, where 1 kge equals 9×10^{16} joules (from Einstein's $e = mc^2$) and is the mathematical product of the shell area and the sun-Earth radiation constant.

5. Just over 60 per cent of the solar helium content, after allowing for the 10 per cent dowry, cannot be accounted for by radiation output. The mass that this represents is absorbed by the newly minted helium as part of its endothermic activity. The expected helium is based on the anticipated 14 per cent of the sun's mass being helium generated by hydrogen fusion during the sun's lifetime. Although the figures given above are based on the hypothesis that the sun's radiant energy output is due to exothermic hydrogen fusion, in fact

hydrogen fusion must be endothermic. But hydrogen fusion, even if endothermic, needs a very high temperature. If the temperature is too low fusion ceases and heat energy from the effects of time slowing on mass would raise the temperature. That is, solar radiant energy is energy released by time slowing, the bulk of that heat energy being absorbed in the process of hydrogen fusion. It is an automatic thermostat.

6. The numbers predict that gravitational energy accounts for 97 per cent of the energy produced as a result of time slowing on the sun's mass. The remainder is used for helium formation, solar radiation, the energy required for Hubble universal expansion, and the energy involved in the sun orbiting the galaxy.

The Nudge Energy Puzzle

The calculations assume that the orbits of the planets are sufficiently circular that small deviations because of slight ellipsis of the orbit can be ignored. However, Mercury has a more marked ellipsis, so that the Mercury-sun distance varies by a small but significant amount so as to render the calculations unreliable for this planet.

The planetary orbits are due to solar gravity. If solar gravity suddenly ceased, the planets would continue their motion as a tangent to the orbital path. This would form a right angle so that simple Pythagorean analysis would show the distance that the planet must be nudged sunwards per second in order to maintain its orbit. The equation describing this is:

Equation Appendix 4.1 $\qquad N = v^2/2D$

where N is the nudge distance in metres and v is the distance that the planet would have travelled in 1 second. It equals the orbital velocity. D is the average distance between the planet and the sun, in metres. The gravitational force between the sun and the planet is given by:

Equation Appendix 4.2 $\qquad g = G \times M_1 \times M_2/D^2$

where G is the gravitational constant, M_1 is the mass of the sun in kilograms, and M_2 is the mass of the planet, also in kilograms. D is the mean sun-planet distance in metres.

Thus the product of the gravitational force and the nudge distance gives the amount of energy expended to maintain the different planets in their respective orbits.

The solar Earth radiation constant is known: 1400 joules per square metre. From this using the inverse square law, the equivalent can be calculated for each planet. The ratio of the sun's gravitational energy output to its radiant energy output is known (see above), so that the amount of solar gravitational energy intercepted by each planet can be determined from the cross-sectional area of the planet.

The nudge distances for the gas giants, because of their huge orbits, were very small; despite their huge masses and the reduced gravitational force (because of their distances from the sun), their large sizes enabled them to intercept enough solar gravitational energy to keep them in orbit. Indeed there was a relative excess of solar gravitational energy available.

For the rocky planets, the situation was markedly different. The calculations show that Mars intercepted enough solar gravitational energy to keep it in its orbit, but the balance was very close. For Earth and Venus, there was a significant shortfall in that the amount of gravitational energy directly intercepted was woefully short. For Earth the amount of gravitational energy directly intercepted was about a fifth of what was required, whilst for Venus it was a tenth of the energy expended in maintaining its orbit. See table below.

Clearly, gravitational energy, carried by gravitons, that should have bypassed the planet is being drawn into the planet, with Venus drawing in the most despite having a smaller surface gravity than Earth. That is, gravitational curvature of space could not account for this difference. It follows that there must be a force arising from the planets which interacts with the solar gravitons, drawing them towards the planet. Such a force was proposed by Burkhard Heim as the sixth fundamental force of nature—he called them interactions. The force arises from within the nucleons of the atoms making the mass. It has been labelled as the gravito-magnetic force. The amount of solar gravitational energy drawn by this force must depend upon the density of those gravitons, but other factors must play a part.

What perhaps are noticeable are the huge amounts of energy expended to keep the plants in orbit, over 2000 kge/sec.

Table 4.A2 The Energy Requirements to Keep the Planets in Their Respective Orbits

	Venus	Earth	Mars	Jupiter	Saturn	Uranus	Neptune	
Mass × 10^{24}	4.84	5.98	0.639	190,000	56,900	8,680	10,400	kg
Distance × 10^8	1.08	1.50	2.28	7.78	14.3	28.7	45	km
Orbital velocity	35	29.8	24.1	13.1	9.66	6.80	5.43	km/sec
Nudge distance	5.66	2.96	1.28	0.11	0.00326	0.0008	0.00033	m/sec
Radiation constant	2680	1400	603	51.7	15.3	3.80	1.55	J/sq. m
Solar gravitational constant × 10^3	264	138	59.6	5.11	1.51	0.376	0.153	J/sq. m
Gravity energy intercepted	338	196	24	908	193	8.8	3.26	kge/sec
Sun-planet g force x 10^{22}	543	351	0.162	41.2	3.65	0.138	0.0678	m/s/s
Force × nudge distance	3420	1150	23	502	13	0.124	0.0025	kge/sec
Surface gravity	7.94	9.80	3.73	24.89	10.40	8.63	11.4	m/s/s

The Temperature of Helium Fission

Air has an average molecular weight of 28.8. The average velocity change with temperature is 0.6 m/sec/degree Celsius. For the same amount of energy a hydrogen atom, atomic weight 1.0079, would have a change in velocity of 3.22 m/s/degree C. For hydrogen fusion to helium, each of the participating hydrogen atoms lose 0.0725 of mass. Einstein's mass velocity equation is:

Equation Appendix 4.2 $M_{new} = M_{old} \times (1-v^2)^{0.5}$

where v is the velocity expressed as a fraction of the velocity of light. That is, at velocity of hydrogen where the atoms lose 0.0725 mass the temperature is 11.5 million degrees.

Helium has a change in velocity of 1.6 m/sec/degree C. From the data derived from helium formation in the sun, the energy equivalent of the mass lost from the four hydrogen fusion atoms represents only 40 per cent of the energy used by helium for its binding energy; the remainder is absorbed from the environment as part of its endothermic activity. But for helium fission all that binding energy must be overcome before fission can take place (or fusion to form larger mass atoms). This means that the helium atom must lose 0.0725 of mass per 4.0026 mass of helium. From the Einstein mass velocity equation, this means that the helium atom must be accelerated to a velocity of almost 19 per cent of the velocity of light. Given helium's relative sluggish change in velocity per degree, this corresponds to a temperature of 35 million degrees. This is just over 3 times the temperature of hydrogen fusion (11.15 million degrees).

Hydrogen fusion, although endothermic, needs a high temperature for the hydrogen nuclei to reach fusion velocity. The energy released by the reduction in mass eventually repays the energy needed to achieve fusion velocity. The energy of H weapons is due to the temperature produced by the nuclear fission triggers being high enough to cause helium fission, despite the losses of energy due to the endothermic consequence of hydrogen fusion.

CHAPTER FIVE

Gravity and Time

Ever since Isaac Newton identified gravity as being an acceleration force, there has been speculation as to the exact nature of gravity. When Einstein proposed his general theory of relativity, and particularly when he proposed that space (or rather space-time) is curved, two theories arose. The first that gravity is a consequence of the curvature of space. That is, following Newton's proposition that a moving object continues in a fixed direction unless acted upon by another force, this has been taken as being a fixed direction relative to space, so that if space curved, the moving body would follow that curvature. This cannot be true. To take two examples, a cannon ball fired from the top of a cliff over the sea will have a terminal velocity greater than the muzzle velocity. Since it takes energy to increase the velocity of a cannon ball, simple curving of space will not supply the necessary energy. Similarly, a falling apple accelerates as it falls. Again, curvature of space will not provide the energy for that acceleration. Sadly, this suggestion has been used in some school textbooks to describe the trajectory of a thrown ball as it curves whilst falling to the ground.

The second theory is that gravity causes a curvature in space. Observations of distant galaxies show that some have a bright beam of light emanating from a collapsing black hole. The beam whilst travelling many thousands of light years through its parent galaxy, show no signs of curvature despite the gravitational field of the billions of stars that make up that galaxy. This raises questions about the extent of the curvature of space in response to the intensity of the gravitational field. Yet the gravitational field of a galaxy will interact with that of a neighbouring galaxy so that clusters of galaxies can form and may even be drawn to each other and collide. It follows that the intensity of the gravitational field must be substantial if it is to have any effect on the light path.

Perhaps it is a matter of scale. Euclidean geometry is based on flat surfaces, and in that geometry the angles of a triangle add up to 180 degrees. Euclidean geometry does have a three-dimensional offshoot called plane geometry. It was this geometry that enabled the builders of the pyramids to ensure that the four triangles that made up the sides of the pyramid met together so that the apex of the pyramid was at a particular height and place.

Curved surfaces have a different geometry, and the angles of a triangle on such a surface could be greater than 180 degrees. Further if the curvature was that of the surface of a sphere, a triangle could have three right angles. Such a situation occurs with the Earth. Two people can set off from the North Pole, one going due west and the other going due south. At the Equator, if they turn a right angle to travel towards each other, they will meet. The triangle of the area enclosed by their travels will have three right angles. Yet one can draw an accurate map, say of Great Britain, using Euclidean geometry. That is, the curvature of the Earth is too insignificant to affect that localised map. It is only when considering very large areas, such as northern Canada, that inaccuracies of mapping will occur when drawing a map based on Euclidean geometry.

The geometry of curved surfaces also has its three-dimensional counterpart. Curved geometry of the interior of a curved body was developed by Riemann, and Einstein used this geometry to develop his general theory of relativity. But curved space presents problems. Binary stars are very common and occur where two stars are gravitationally bound to orbit each other. If each star's gravitational field curves in space, there must be some point where the curvatures are in opposite directions. This means that a beam of light must develop an S-bend. More importantly it implies that space is a material and one with a variable density. The only resolutions of these difficulties are that if gravity does cause space to curve, it is a very localised phenomenon and that overall for the whole universe, Euclidean geometry (which appertains to flat surfaces) prevails, or in the parlance of astrophysicists, the geometry of the universe is flat. Current observations have shown that it is at least 99 per cent flat.

Einstein's general relativity theory leads to the prediction that starlight would be deflected by approximately 1.7 arc seconds if it passed close to the sun, where an arc second is 1/3600th degree. This was confirmed both photographically and more recently by long range radio interferometry (a system of radio astronomy when radio telescopes are separated by a very long distance; this improves the resolution to that of the best optical telescopes). The resulting accuracy is now approximately within 1 per cent.

The theory also predicted the orbit of the planet Mercury. This planet has an orbit which is much more elliptical than any other planet. Moreover the orbit precesses. This interpretation has been challenged using Burkhard Heim's calculation of two additional fundamental forces. One coincides with the acceleration force described in earlier chapters as being responsible for the Hubble constant, the other has been called the electromagnetic gravitational force. This force has recently been reproduced in the laboratory. It has been suggested that this force could account for Mercury's orbital pattern.

It is possible that there may be other causes for the bending of starlight by the sun's gravity. It is stressed that this is a hypothesis and that although the underlying principle is valid, it may not account quantitatively for the extent of the observed deflection of starlight by the sun. This is refraction of light caused by gravity.

According to the special theory of relativity, when the gravitational force equals that of the velocity of light, then within a moving body the second is infinitely prolonged. Since the velocity of light is a time-dependent fundamental constant (that is, it is constant in all time frames), if the second is prolonged infinitely, the velocity as far as the photon is concerned has become zero and so is not moving. But to an outside observer the photon is moving. Yet by definition photons cannot escape from a black hole. Two things follow. When gravitational acceleration is much less than the speed of light, there will still be some time slowing, the extent of which will depend upon how strong that gravitational force is. In Earth observations where gravity is approximately 10 m/sec/sec, there is no time slowing and there is no curvature of a ray of light. The second thing is that to an outside observer photons must move. They are not stuck on the event horizon like so many flies on flypaper. They cannot travel in reverse—their structure as waves with a uniform density of energy—that energy is the minimum, the quantum of energy it cannot be reversed. Photons however can be refracted, that is turned to travel in another direction. At the extreme the turning can be 180 degrees so that the photons crash into somewhere close to their point of origin. That extreme depends upon the strength of the gravitational force.

Refraction occurs when a beam of light strikes, at an angle, the surface of something transparent and in which the velocity of light is slower than the velocity of light in air. The velocity of light in water is half that of the velocity of light in air. The light travels as waves. When that wave strikes the surface of the water, that part of the wave in the water will have its velocity slowed before the rest of the wave. As a result the wave form is swivelled until all the wave is in the water. Then it continues in a straight line. This effect can be readily seen by looking at a stick stuck vertically in the mud of the bottom of a clear lake. The stick appears to be crooked.

There is an analogy of a skier travelling downhill. When the skier wishes to make a turn, he transfers some of his weight on to the inner ski, increasing the frictional drag on the under surface of the ski, so slowing it. The outer ski has less friction under its surface and travels faster. If the skier wishes to increase the sharpness of the turn, he increases the pressure on the inner ski, increasing the drag. Then as he comes out of the turn, he eases the downward pressure on the inner ski so that the turn becomes less severe. Finally he balances his weight evenly on the skis and so continues in a straight line. But the key point is that the bow front of, say, his face and helmet continue in the new direction.

Equally, if the wings of a wave of light are forced to travel at different speeds, the wave will turn until both wings of the wave are travelling at the same speed and the bow front continues in the new direction (i.e., if gravity has caused the different speeds, then gravity is inducing refraction).

The gravitational force at the surface of the sun is 272 m/sec/sec. This is more than 27 times the force of gravity on Earth and would be enough to kill a person. At a distance of 5 solar radii from the centre of the sun, the gravitational force is approximately that of Earth—whose gravity does not deflect light in any detectable fashion. Thus the line of a light beam from a lighthouse and its straightness does not alter irrespective of which way the beam is pointing. Any incoming light wave with a light path that will pass fairly close to the sun's surface will experience solar gravity at a strength equal to that of Earth when the approaching wave is fractionally fewer than 5 solar radii distance from the surface of the sun. If the wave passes closer still to the sun, the wave will be influenced by solar gravity until it has travelled another approximately 5 solar radii from the sun. If the wave

Chapter Five

grazes the surface of the sun, it will be experiencing a gravitational force that is greater than that on Earth for around 22.5 sec. During that transit the gravitational force would be steadily increasing and then steadily decreasing—like our skier making an increasingly sharp bend and then coming out of the bend.

The extent of the bend will depend up how much gravity slows time and the width of the wave. When the gravitational force is creating an acceleration of 272 m/sec/sec, the second is expanded by 0.45 pico-second. This is too small to be of significance.

One can take this refraction concept further. In 1918, there was an eclipse of the sun whilst the sun was in front of a fairly rich star field. Photographs of the eclipse were taken in Brazil, and the changes in position of the stars were used to calculate the extent of the deviation of the beam of star light from the various stars if the beams grazed the surface of the sun. (More details can be found in the book entitled *The Golem* by Collins and Pinch, 1997, Cambridge University Press.) The mean, taken from eight photographic plates, was calculated as 1.98 arc seconds, where 1 arc second is 1/3600th of a degree. Einstein's theory of general relativity predicted a value of 1.75 arc seconds. The observations had a statistical scatter such that there was a probability of only one in ten that Einstein's theory and calculations resulting from that theory were right. Nevertheless, it was taken as proving the correctness of the theory of general relativity theory. At the time there was no concept that the fundamental constants and their forces could interfere with each other. We now know that there are interactions between these fundamental forces (e.g., gravity and the expansion constant described by Burkhard Heim, the heat of expansion of the sun, their photons against gravitons that cause gravimetric compression so keeping the radius of the sun constant).

Taking the 1.98 arc second calculations as being the true effect of solar gravity and the solar gravitational acceleration being 272 m/sec/sec, then the strength of the gravitational force to induce a 180 degree turn of a photon is $\sim 9 \times 10^7$ m/sec or approximately one third of the speed of light. That is, photons are turned to travel in the reverse direction to crash back into the substance of a black hole before the graviton has fully emerged from its source. Since the graviton has a wavelength of perhaps several kilometres, as it starts to emerge the energy density of the emerging tip will be low. But as emergence progresses the energy, which is coming from the source, increases. Furthermore gravitons are compressed by the sheer density of numbers so that during emergence their shape becomes ever more javelin-like. This concentrates the energy density of the tip so that it is sufficiently strong to tip (refract) the fleeing photon. As more and more of the great graviton wave emerges, the energy density at the tip increases, giving yet more force to refract the photon. By the time enough of the graviton has emerged so that the energy density being generated has a force with an acceleration that is a third of the speed of light, the photon has been refracted 180 degrees and so is travelling back to the surface of the black hole. That is, the event horizon of a black hole is about one third of the wavelength of a graviton above the surface of the black hole.

There are thus two possibilities to explain the apparent deflection of a beam of starlight as it passes the sun. The first is Einstein's proposal that around a body that has a high gravitational energy output, space becomes curved; the second is refraction. Which possibility is correct could be tested by checking whether radio waves passing close to the surface of the sun suffer the same deflection quantitatively as white light. If they do then Einstein's interpretation is correct. If there is a difference then refraction must be considered, as the degree of refraction depends upon the wavelength of the electromagnetic input.

The Wave Particle Dilemma

Photons can be considered as particles with a small volume, or as waves. Their behaviour depends upon the circumstances. White light has a series of wavelengths, all close to each other. In a non-polarised beam the forward orientation of the wave can be in any direction. If the light wave or photon has to pass through a narrow slit and the bow front can just squeeze through the slit, the momentum of the bundle of energy that is the wave will cause its "wings" to be drawn in (the wave or the energy that forms the wave cannot fracture or get stuck).

The analogy is that of the gannet, a sea bird with a wide wing span which dives into the water to catch fish. As it approaches the water surface it folds its wings tightly against its body, minimising the drag once it is under water, enabling the bird to catch fish at a depth. After the light wave has passed through the slit, as the two slit experiment shows, the waves unfold, causing the stripes seen when the light from the two slits reaches the display surface.

Refraction

Refraction depends upon the refractive index of the material (or how fast light travels through the material) and the width of the incoming waves. Light consists of a mix of the colours of the rainbow, with each colour having its own wavelength and therefore its own wave width. Hence for a lens there is chromatic aberration unless the lens has been especially designed and created to compensate for this.

Light waves can summate. This increases their amplitude and their wave width. Light from a distant star, which is destined to graze the surface of the sun, starts as an extremely dense quantity of photons. Summation must occur. As the photons move from their source they spread out, and as the light is non-polarised their wave width increases. As the edges of the wave touch the edge of adjacent wave, energy will flow into the wave edge that has the lower energy density. Logically the waves must join up to form a composite wave which steadily expands as it moves farther and farther from its source. The composite wave must be easily fractured if the wave nudges a lump of mass. On arriving near the sun's gravitational force the composite wave enters a zone where the pace of time is very slightly slower. The wave starts to refract. The outer edge of the wave will be swung forward until its velocity equals that of the velocity of light applicable to that particular location. Any further refraction and the composite wave will start to crumble at the far side. Meanwhile the composite wave continues entering territory where the force of gravity is increasing, and with it, slightly more time slowing, causing slightly more refraction. This will continue for up to ~22.5 sec until the wave is out of the sun's gravitational maw.

Quantitatively it is not possible to gauge the extent of the refraction, as the wave width of the photons that emanated from such a long distance cannot be determined. It is emphasised that the concept of a composite wave is very much a hypothesis, although a logical one.

This is not to decry that the mathematics of general relativity theory explain the deviation of the light as it passes close to the sun. However, it is that just the interpretation of or the conclusions drawn from that mathematics that may be questionable. There is already one example where one conclusion drawn from special relativity is proving false (that mass increases with velocity when the opposite must be true if the laws of physics are constant for all observers, as was discussed in an earlier chapter).

The phrase "curved space-time continuum" implies that space is a material that is apparently capable of infinite stretching and, more significantly, that time and space are physically entwined. The theory assumes that the period of the second is constant apart from any expansion due to velocity. Yet the period of the second cannot be constant if the Hubble expansion constant is constant. A more cautious approach is that general relativity theory predicts changes by an as yet ill understood process that simulates both that space is curved by gravity and that time and space are interlocked in some form of continuum. It follows that with the discovery of the expansion of time the whole mathematics behind the theory of general relativity may have to be re-examined.

How Gravity Operates

Gravity is a strange force of acceleration. The acceleration it produces follows the inverse square law, but when striking a body, unlike light, it is not affected by the presenting area of that body. Neither is it affected by its mass. That is, the resulting acceleration is the same irrespective of mass. Yet the greater the mass, the more energy is required to accelerate that mass. Two masses of the same size, shape, and orientation, but with different densities, with both at the same distance from the ground, will experience the same acceleration. The extra energy required for the acceleration of the more dense mass must come from the interaction between gravity and that mass. This is particularly apparent when considering very large masses, such as planets, which are considerable distances from their primary source of gravity. This is even more apparent when considering binary star systems, which mutually orbit each other like two ice dancers in a spin. The distances between the two stars must mean that the contribution of gravitational force received by the recipient star is very small. This is even more apparent when considering the acceleration towards collision between two galaxies that are millions of light years apart. Somehow the energy from gravity is changed to reinforce the acceleration particles on the surface of the nucleon. Gravity provides the energy for the work done in accelerating mass.

Gravity is mediated by gravitons. That is, they are particles of energy that, like light, travel as waves, but given the wavelength of a gravity wave is several kilometres, each wave probably contains many gravitons. The only logical explanation is gravity waves interact with each and every nucleon in the mass, no matter how many nucleons there are within any atomic nucleus. Further, the interaction must be the same for each nucleon. The total effect, that is, the extent of the acceleration, must depend upon the density of the gravitons per unit volume as that volume comes into contact with a mass. Another oddity about gravity is that though gravitons must possess momentum, the effect they produce is in reverse direction of that momentum. Gravitons must also be extremely narrow to pass through the dense nucleus of the atom. Some gravitons will hit a nucleon (and eventually all the nucleons will be hit) with enough surviving to reach the furthermost nucleon within the nucleus, even if that nucleus is packed with over a hundred nucleons. Yet more gravitons will pass through the atom unscathed to penetrate the next atom in their path.

Burkhard Heim's fifth fundamental force, the one underpinning the Hubble constant, calculated that this acceleration force interacted with gravity to the mutual inhibition of their respective forces. The acceleration force must be emitted from all the surfaces of a nucleon. It has a very short range so that if the nucleon is moving, the prow of that nucleon will absorb the acceleration force particle at the prow, leaving the hindmost surface of the nucleon with a force that is unopposed. Otherwise the acceleration force is primarily a compression force, as the acceleration force particles dotted on opposite sides of the nucleon cancel their efforts. Gravitons inhibit the acceleration particles that are on the face being struck by the gravitons. The inhibited acceleration particles quickly recover, only to be assaulted by yet more gravitons as another gravity wave follows. Meanwhile the transiently unopposed acceleration particles on the opposite side of the nucleon accelerate the nucleon. In this way the body containing the nucleons is accelerated towards its target.

The change in acceleration is enormous. Using Newton's fabled apple story, before the apple broke loose from the tree, it was subject to an acceleration force of 5×10^{-10} m/sec/sec. But there was no movement relative to the tree, and so no work was done. When the apple started to fall, it was accelerated to 10 m/s/s. That is, work was done and energy expended. This sudden 20 billionfold increase in acceleration occurred almost instantly for each of the nucleons in that apple. That is, the gravitons from Earth must have inhibited many acceleration particles on the proximal, that is Earth facing, surface of the apple, releasing the distal acceleration particles full range. Since energy cannot be created or destroyed (although it can change its character, e.g., become mass) the energy released when the proximal acceleration particles and the gravitons annihilate each other must pass through the nucleon to fuel the acceleration particles on that distal surface. Some energy must be retained to re-form the acceleration particles on the proximal surface of the nucleon. The alternative possible explanation is that the energy released would create heat or light, but this is not what is observed.

When released from a large mass such as the sun, gravitons must be extremely narrow and very densely packed. As they radiate out from the sun, the density of gravitons limits their sideways expansion. But gravitons criss-cross each other, the sun's smaller gravitons passing through the expanded gravitons from distant large bodies (e.g., those from black holes). But the whole of space is filled with gravitons, so that there is no location in space that is not influenced by gravity to some degree. Whether gravitons flow in giant waves like a dense flock of birds wheeling through the sky or are merely a torrent still awaits experimental determination.

When the flood of gravitons from the sun reaches a large body, such as Jupiter, many will pass through but many also will encounter a nucleon and effectively be destroyed. An analogy is a tidal wave approaching a beach that meets a cluster of rocks. Many of the droplets of water that make up the tidal wave will be lost, thrown up into the air as a great spume. But the wave continues; much water passes between the rocks and emerges to be joined by water flowing in from the sides, water that has not struck the rocks. Thus the great tidal wave continues beach-ward to crash as giant surf of more or less equal height along the beach. Gravitons, when expanding at their sides, dilute the energy density within the gravitational wave. The sideways expansion is limited by a similar expansion of the adjacent gravitational waves. If the expansion wings are of equal energy density, then they will stay in the same formation. But if the energy density of one is diminished, the adjacent wave will expand. This will create a domino effect along the whole gravitational wave front. When emerging from a cluster of nucleons, there are gaps; those gravitons that get through will start to expand, but as their energy density is less, they will be compressed by the expansion of adjacent gravitons until all are at equal energy density. The expansions fill the gap—much like the tidal wave fills the gap of the water lost hitting the rocks.

Gravity casts no shadow. This also means that as well as travelling outwards from the sun, gravitons can also shuffle sideways, filling the gaps caused from the loss of those interacting with the nucleons. This recruitment of gravitons from the side explains why the acceleration of an irregular mass is not affected by the orientation of that mass. A spear-shaped asteroid will have the same gravitational acceleration whether it is approaching another heavenly body horizontally or vertically.

APPENDIX 5

The Effects of the Expansion of Time on the Fundamental Constants and the Basic Units of Measurement

The Hubble expansion constant has a value of 5×10^{-10} m/sec/sec. If the second is more brief compared with Earth time, the value of the constant will increase. That is, the acceleration will be constant in all time frames or changes in the pace of time. The acceleration will therefore be in relation to the prevailing pace of time. That is, the Doppler shift (on which the calculations of the velocity and so acceleration are dependent) will be in relation to the velocity of the moving particle, but that velocity will in turn be in relation to the pace of time of the particle's environment. To an independent static observer on Earth in a different time frame, the acceleration will be increasing, as would be shown if he could use an independent means of measuring velocity.

But an increase in acceleration requires an increase in force. Force is measured in Newtons, that is, the faster the time, the greater the "strength" of the Newton. It follows that the Newton is constant in all time frames. The Newton by definition is the product of mass and acceleration. If both acceleration and the Newton are constant in all time frames, then mass must be constant relative to the prevailing time frame. It follows that the faster the pace of time, the greater the mass. Conversely the slower the time, the less is the mass. Since mass is concentrated energy, it is this, via the universal slowing of time, which provides the fuel for the expansion of the universe and the energy of the sun and all other stars. Gravity is also measured in Newtons. It follows that gravity is constant in all time frames. The faster the time, the greater the gravity. Gravity rests upon the gravitational constant.

This constant must therefore be constant in all time frames, just as one would expect for any fundamental constant. But the gravitational constant in turn is a product of Newtons and metres squared divided by mass (kilograms) squared. If the gravitational constant is constant in all times, together with both mass and the Newton, then length, metres, must be constant in all time frames. As time goes faster, so the length of the metre will increase with respect to the Earthbound metre. Conversely if time slows, the metre will shrink in length. This is in accord with Einstein's special relativity theory.

Distance as opposed to length will depend upon the units by which it is measured. As far as astronomical distances, these are measured either in parsecs or light years. Light coming from a very distant supernova will be in a faster time frame. The metres it is travelling will at first be prolonged as compared with the standard Earth metre. However, the year, relative to the Earth year, will be much shorter. The combination therefore cancels out. The light year, therefore, when calculating distances to distant supernova is unaffected by any changes of time. What applies to light years also applies to light seconds and to fractions of the light second. If these are unaffected by changes in the pace of time, this seems to contradict the relativity conclusion. But the stability of the light year as far as distance is concerned is essential for the calculation of the Hubble constant and its derivative, the Hubble acceleration constant.

It follows that if distance is stable, the inverse square law is applicable to the determination of distance. The Hubble constant is based on observing the magnitude ofa distant supernova which are exploding at different paces of time. The concept of the supernova acting as a standard candle is that the light decays at a standard rate. That is, it is time dependent. If the time is faster, then the rate of photon emission must be greater with respect to Earth time. Brightness or magnitude should be affected, but only if the amplitude of the light waves are unchanged by time. The amplitude or brightness of a light wave depends upon the speed of light—thus the speed of light in water is half that of the speed of light in air. Underwater colours appear much brighter, but their frequencies are unchanged. That is, the wavelengths are shortened but amplitude is increased. Equally when time is faster, the speed of light is faster, and therefore the wavelengths of the light from the supernova, as they are emitted, are stretched, reducing their amplitude. Brightness, or magnitude, therefore must consist of the combination of both amplitude of the light waves and the frequency of photon emission. It is only with this combination that magnitude becomes a reliable measurement.

Length, mass, acceleration, velocity, and the Newton are constant at all paces of time, but the unit of energy (Newton per metre) is independent of any change in the pace of time, as the effects of time cancel out. It follows that changes in mass with changes in the pace of time require either the release or the addition of joules. If length varies with the pace of time, then wavelength must also vary. The faster the time, the longer the wavelength; this must mean that frequency, like the joule, is unchanged. This has important relevance when considering the frequencies in light from very distant galaxies which originate in different time frames. The Doppler shift therefore is unaffected in transit when the light from a very distant galaxy passes through progressively slower and slower time frames in its journey to Earth, although the actual Doppler shift will depend upon the velocity of the moving galaxy (or supernova) and the time frame of the environment when the light (with its Doppler shift) was emitted.

The Relativity Contradiction

Einstein's special theory of relativity proposes that mass increases and length decreases with velocity. This creates considerable difficulties when considering such things as the gravitational constant, which by definition is constant in all time frames. The special theory also says that the laws of physics are the same for all observers. This contradiction is best shown by a worked example.

Einstein's error: Einstein's mass velocity equation was derived from the calculation of the changes in momenta of objects thrown from a moving platform to a stationary catcher and vice versa. The calculations showed that the momentum and so the mass of the object coming to the stationary catcher's arms was apparently greater than when it was with the thrower on the moving platform. It was then a question of interpretation. Einstein thought that increase in mass was occurring at the throwing platform. The following thought experiment shows it must be otherwise.

Two friends agree to throw weighted rubber masses towards each other so that the masses collide with the same momentum. One is stationary and is a short distance from a railway track, the other is on a passing train travelling at a relativistic velocity such that its time is expanded, that is slowed, by a factor of two. The stationary man throws an eight kg ring towards the train so that it is travelling at 10 metre/sec. The man on the train sees the ring coming towards him but to him it travels 20 metres in his second. He must therefore throw his ring at 20 m/s. But if the momentum on ring collision is to be the same he must use a four kg ring. (Momentum is mass x velocity.) The ring collide and bounce back to their respective throwers. That is to achieve the same momentum mass must be reduced if time is slower. The oddity is to each of throwers the velocity of the two rings were the same but neither pair would ever agree as to what that velocity was. More significantly the slowed rate of graviton emission (which determines the mass) from the mass on the moving platform and potentially detectable by the stationary person would indicate that the mass was less. Regrettably the view that mass increases with relativistic velocity is still widespread. As is discussed more fully in Chapter 7 the reduction in mass with increased velocity provides the energy for Jupiter's gigantic equatorial storms.

Equation Appendix 5.1 $\quad\quad Mass_{new} = Mass_{original}(1-v^2)^{0.5}$

This equation, showing that mass reduces at high velocity, gets around the conflict posed by the question of how fast a pendulum would swing if it were on the moon and the moon orbited the Earth at high velocity. For the pendulum the slowing of time produced by relativistic time dilation would make the pendulum swing slower. The high velocity would, according to Equation Appendix 5.1, also reduce the mass of the moon, thereby reducing its gravitational force on the pendulum bob, and the effect would make the pendulum swing more slowly.

Quantifying Gravitational Energy

Since time started, it has doubled 19.1 times. Since mass is inversely related to time, mass has been reduced by the same factor. The loss of mass is shed as energy, with each kilogram loss (kge) worth 9×10^{16} joules (from $e = mc^2$). Each doubling period is 8.84 billion years of cosmological time. Putting it all together:

Equation Appendix 5.2   Energy loss (kg) = Mass (kg)/
                             Cosmological age (seconds)

If the mass is 1 kg, the energy released by time slowing is:

Equation Appendix 5.3   Energy = $1 \times mc^2/(5 \times 10^{18})$ joules
                             = $c^2/(5 \times 10^{18})$ = 0.018 joules/sec/kg

As was shown in the previous chapter, 97-98 per cent of the energy released by time slowing is gravitational energy; therefore, the gravitational energy released is 0.0176 joules/sec/kg of mass. This rate will change with time, as time is expanding exponentially, but the rate of curvature of the curve of time, that is plotting the pace of time against the duration of Earth time, has become so flat that currently this figure will not change significantly for another million years.

CHAPTER SIX

The Earth and the Moon

This chapter shows what catastrophic effect can result when the effects of time expansion in conjunction with other energy providing mechanisms act on a small body. The chapter also provides two separate proofs of the effects of the expansion of time, effects that can be observed today.

If it were possible to enclose 1 kg of ice in a heatproof container so that no heat could escape, and leave it for 2 billion years, the mass would have been reduced by a factor of 2/169 where 169 is the age of the universe in billions of years of cosmological time. That is, the kilogram of ice would have lost 12.37 g of its mass as energy as a result of the expansion of time during that 2 billion years. Most of that energy would have been gravitational energy, but 2 per cent would have been as heat energy. This corresponds to 0.25 g of water, or 2.29×10^{13} joules of heat energy. The heat capacity of water is 4200 joules/kg per degree Kelvin. If none of the heat escaped, the 2.29×10^{13} joules of heat energy would raise the temperature by 5.5 million degrees, more than hot enough to vaporise the ice and cause the water molecules to dissociate into hydrogen and oxygen. The internal pressure within the container would be immense, enough to start splitting the container. The sudden drop in pressure as the gas mixture started to escape would lower the temperature into the range where the two gases could associate again and re-form steam. Any spark would cause the gas mixture to explode and to explode very violently.

This is what happened, and happened on a planetary scale, with dramatic consequences for the Earth.

Between Jupiter and Mars, there probably was a rocky planet. Its size is unknown but most probably it had the same mass as Mars (6×10^{23} kg). In its formation it accumulated large quantities of ice, just as Mars and the Earth did. Just like the other rocky planets, it had some radioactive elements, and this would have heated the interior so that the ice there melted and was squeezed towards the surface. But the surface was frozen. Unlike Earth and Mars, it was too far from the sun to benefit from solar radiation, so a thick crust of ice mixed with rock formed on the surface, forming a substantial heat barrier. Meanwhile the expansion of time released heat deep inside the planet. A mass the size of Mars being ten million times smaller than the sun, would generate 100 kge of heat energy per second. Initially little if any of that heat could escape. Over the first couple of billion years that this planet was in existence, the heat built up, raising the internal temperature higher and higher until the ice inside, and particularly the underside of the thick ice bound surface crust, melted and then boiled. The heat production continued, but the ice-toughened crust resisted expansion. The temperature rose so much that some of the high pressure steam deep inside the planet dissociated into hydrogen and oxygen. Eventually the whole internal pressure was so great that the planet began to split, and so the stage was set for a huge explosion as the escaping gases enabled cooling. The resulting explosion shattered the planet. Its residue is the asteroid belt.

Fragments from the shattered planet were thrown in all directions. Some of those thrown outwards were captured by Jupiter to become rocky moons orbiting that gas giant. (The presence of rocky moons orbiting Jupiter and Saturn is anomalous, as this is deep in gas territory, and the gas giants in their formation would have swallowed up any rocky material in their orbit.) A large number of fragments continued in the same orbit, forming the asteroid belt. A substantial number of fragments were thrown towards the sun, but their angle of motion relative to their orbital path was sufficiently low that they formed a cloud of debris that embarked in a very prolonged death spiral towards the sun, taking perhaps a million years or more to get as close to the sun as the inner planets. There they left their marks during a period of intense bombardment on Mars, the Earth, and the moon, as well as Venus and Mercury. It should be noted that the asteroid bombardment could only have arisen from within the solar system. Any possible alternative source is too far away. But eventually the fragments that missed these bodies spiralled further into the sun. The bombardment was over.

Other fragments were sent directly towards the sun and travelled there at the speed of Halley's comet to smash into the sun. But a number of fragments from the shattered planet had sufficient orbital momentum

Chapter Six

to be displaced towards the sun but at a slight angle. Their paths were such that they missed the sun only to loop around and be slingshot by the solar gravity to form long range comets. (There is no other satisfactory explanation for the relatively narrow orbital paths of these long range comets.) Solar gravity ensured that the velocities reached by these comets would be similar to those reached by Halley's comet.

In the time that Halley's comet crosses the Earth's orbital path, loops around the sun, and emerges to cross Earth's orbital path again, it has travelled some three Astronomical Units (AUs, the distance between the Earth and the sun). It does this in a few months. Earth takes a year to travel a little over six AUs as it orbits the sun. That is, the average velocity of Halley's comet when it is inside Earth's orbit is approximately twice that of Earth's orbital velocity. At its peak at its closest approach to the sun, the velocity would have doubled again. That velocity would not suffer any deceleration until it had well passed the sun on its way back to the Oort clouds. Similarly an asteroid from the shattered planet would have followed a similar path and so would have a very similar velocity as Halley's comet until it hit Earth head on.

The alternative scenario is that the asteroid belt is the fragmentary debris from the giant supernova that spawned the solar system. The explanation currently given is that gravitational perturbations from Jupiter prevented the debris from forming into a planet. There are problems with this scenario. Unless the asteroids had already formed a full ring, they would be close to Jupiter for only a few months every fifteen or so years. More significantly, at the time of the supernova, individual atoms were formed and expelled at high velocity by the radiant pressure of the supernova. As the atoms were drawn to the nascent sun, they would combine, and thereafter it is an exponential process of pulling the atoms and molecules into masses, which in turn would gravitate towards each other. Many of the asteroids are large, some very large. Any perturbation would have its greatest effect when the would-be asteroid fragments were very small. Some collisions were inevitable, but the impact velocities would be comparatively small, too small to push any one large asteroid sunwards. This scenario provides no explanation for the great meteorite bombardment that so pockmarked the moon, nor for Jupiter's many rocky moons.

Chapter Six

The consequences for the Earth and the Moon

The Earth was formed from the debris of a giant supernova, which also spawned the whole solar system. The sun quickly condensed into a star and gravitationally gathered around it all the debris from that supernova. Most of the heavier and denser masses fell into the sun, but enough remained to orbit the sun and gravitationally collected into four (but almost certainly five) rocky planets, each being at a different orbital distance from the sun. The two outer rocky planets were Mars and probably another small Mars-like planet midway between Mars and Jupiter.

It took 1.5 billion years for the debris to be assembled to form planet Earth. That is, the sun is an estimated 6 billion years old, whilst the Earth is an estimated 4.5 billion years old. During its formation, the proto-Earth became very hot, with an internal heat production up to 60 times its present rate due to its relatively high radioactive content. The half-life of these radioactive elements is about a billion years, although it varies with different radioactive elements as well as varying with different isotopes of those radioactive elements (that is, tracing backwards, the radioactive energy output doubled every billion or so years). In addition there was, on Earth, the constant production of heat from time expansion, which was adding another 11,250 kge/ per second. (A kge is the energy equivalent of 1 kg of mass and equals 9×10^{16} joules, from Einstein's famous equation, whilst 4.18 joules is the energy required to heat 1 g of water by 1°, that is, 1 kge of energy would boil approximately 20 million tons of ice.)

By the time the Earth was stabilised it was very hot with a thin crust over molten magma. There were no mountains, or at least no big mountain ranges, as the crust could not support their weight. Equally there were no deep oceans. The plasticity of the semi-molten internal Earth ensured that the Earth was symmetrical and rotating whilst it orbited the sun. The atmosphere was a mixture of nitrogen and water vapour. There was very little carbon dioxide as any high concentration of carbon dioxide would have triggered a greenhouse effect and would have caused the sea to boil, particularly as the atmospheric pressure was four fifths of its present value. The sea consisted of a single ocean that covered the Earth to a depth of around 2.5 km. (A standard reference atlas describes that the oceans cover approximately 70 per cent of the world's surface and have an average depth of 12,000 ft. If this volume of water was spread uniformly across the globe the average depth would be approximately 2.5 km.)

Then it happened.

When the Earth was between 500 million and 1 billion years old it was struck by a huge meteorite which caused the Earth to lose 1.23 per cent of its mass. Ten billion billion tons of Earth's mass was thrown with sufficient force that the velocity exceeded the Earth's escape velocity (11.2 km/sec).

The following is a reconstruction based on the geophysics of the Earth as well as the astronomical history of the inner solar system. It is based strictly on proven physics and is what must have happened when a large asteroid struck the Earth to create the moon.

The impact created a huge tsunami which radiated out in all directions with the waves reaching a height of at least 15 km. The tsunamis raced around the world and within hours collided and lost velocity. The water flowed back towards the crater, but with a slope of around 1:750 it would have taken many days to get back to the crater.

Meanwhile, a huge amount of mass was thrown out. Approximately 1.2 per cent of the mass of the old Earth was ejected at a velocity that exceeded the escape velocity from Earth. A reasonable approach is that a similar volume did not reach escape velocity but was thrown up and outwards into the air to fall back to Earth a distance away from the crater. A speculative possibility is that Ayre's Rock in the middle of the Australian desert was the result of part of that upheaval—its geology is so different from its surroundings.

The crater was enormous and covered an area approximately equal to the sum of the areas of both the Pacific and Atlantic oceans. (The Americas at that time were attached to the Euro-Asian continental mass and to Africa.) The edges of the crater are marked by the edges of the continental shelves where the bottom of the ocean suddenly plunges very steeply to great depths. If the ejectate was virtually only the mass that formed the moon the maximum depth of the crater would have been around 300 km. But if the volume of the ejectate was twice that of the moon, the crater would have had a peak depth of more than 500 km. Molten magma would have poured in from the sides of the crater, but given the sheer size of the crater, this would have taken a very considerable time to fill. The most fluid part of the magma would have been at the greatest depth, and so the lowest part of the crater would have filled in more quickly.

At the upper edges of the crater, magma would have flowed more slowly but would have undermined that portion of the crust surrounding the crater, and that too would have collapsed and been dragged into the crater, carrying with it sand and sandstone. These would eventually have profound effects on the Earth and indeed on the development of life. But magma would have been drawn from below the Earth's crust from all around the world. The Earth's crust would have been undermined and would have collapsed. This would have led to shrinkage of the planet. With the loss of over 1.2 per cent of the Earth's mass, the land level would have fallen by 43 km (see appendix to this chapter). The crust would have buckled, raising mountains and creating great elevated plains—such as those in China or the Americas.

The huge crater left exposed a gigantic lake of raw lava at temperature in excess of a thousand degrees. The lake would have had an area of around 250 million square kilometres. Its surface temperature would have been about 1/6th that of the sun. The initial radiant heat energy would have been around 30,000 kge per second. This was vastly in excess of what Earth was generating, even allowing for the increased energy (compared with today) from radioactive sources. The amount of heat escaping would have created an immense updraft, and this would have created huge storms in the Earth's atmosphere. Winds of many hundred miles per hour would have been generated. The initial amounts of water flowing back and then tipping over the edge of the precipice, which was the crater's edge, would have crashed into the lava lake, penetrating it deeply and boiling as superheated steam. The steam would then have risen with the updraft high into the atmosphere, where it would have cooled and condensed. The whirling air currents around the updraft would have carried huge clouds, which would have rained as huge downpours and violent storms, creating great rivers, which would have further eroded large parts of the upper crust; that is, more sand and sandstone would have been washed into the crater. Of equal importance, this would have also washed out a lot of salt, as hitherto the ocean would have had little to no salt dissolved in it.

The storm clouds themselves would have generated huge lightning storms, which could have flashed to the Earth. Stagnant or semi stagnant pools would have received sufficient energy to form a variety of chemicals including amino acids, in effect creating a primeval soup.

Water would still have been flowing back into the crater, but the loss of so much heat energy at a rate greater than the Earth could generate, together with the cooling effect of so much rain, meant that the surface crust cooled and thickened, and thickened very considerably. The effect of this would have strengthened the crust so that it held in one piece. It would also have reduced the flow of magma into the crater as the crater half filled.

The cooling effect also would eventually have resulted in much of the Earth being covered by ice, creating a Snowball Earth, except for the large hot ulcer that was the crater. But even that eventually became covered with ice.

Snowball Earth would appear to have lasted over 2 billion years. The evidence for this lies oddly enough in biology. All multicellular organisms have extracellular fluid, that is, fluid that lies between cells, such as plasma, cerebrospinal fluid, joint lubrication fluid, sap, and so on. This extracellular fluid has a salt content of 0.9 per cent. That is, multicellular forms began when the oceans had a salt content of 0.9 per cent. For the cells, with few exceptions, a salt solution that differs from this 0.9 per cent value results in the death of the cell. Cells have evolved strong mechanisms to maintain an interior ionic concentration compatible with this 0.9 per cent saline concentration. But multicellular organisms did not evolve until around the start of the Cambrian period, which was only 600 or so million years ago. Since then the salt content of the oceans has risen to 2.8 per cent. All that additional salt has been washed into the seas by rivers. The inference is that there were no major rivers until the Earth was 3.5 million years old. Before then the bulk of the Earth's water was frozen, covering the land with ice, much like Greenland. It was truly Snowball Earth.

There was one benefit from this massive ice sheet. When the great meteoric bombardment occurred, the one which so pockmarked the moon, the ice sheet absorbed the impact. When that ice eventually melted, there was no trace of the great bombardment. The alternative theory, that the evidence on Earth has been destroyed by weathering, is too simplistic. If evidence of ancient global glaciations can be found in tropical zones, and it can, then weathering could not have been so totally effective. There should be evidence of the bombardment if it ever reached the land surface. Very old and hard rocks, such as the Laurentian Shield in Canada, should show some residual effects.

Another effect of the Earth losing more than 1 per cent of its mass from one side of the globe meant that it was unbalanced. Its rotation would then have precessed, placing great stresses on the crust. The great super continent would have felt a dragging effect at both its eastern and western edges as the under-structure was being dragged towards the crater. The residual effects of this imbalance can still be detected as a small wobble of the Earth's axis of rotation.

Filling of the crater would have ceased when the level of the surface of the lava lake was at the height of the under-surface of the cold, thickened crust. There things stayed stationary. The surface of the lava lake would have cooled and solidified and been covered, like most of the world, with ice. But the Earth was very slowly building up its internal temperature again. The radioactive elements had decayed sufficiently so that heat from this source was now no more than 4 times its present rate of heat production. The expansion of time would have generated an additional several million kge of energy. The temperature of the inner Earth very slowly rose, and with it, some melting of the under-surface of the thickened crust occurred, allowing more magma to flow into the crater. It would take almost 3 million years before enough heat had been generated to melt and thin the crust to its present depth. It is not surprising that the thinnest part of the Earth's crust is under the Pacific Ocean. For more than 2 billion of those years, the Earth would have been an ice ball, trapping all the generated heat inside.

As the crust thinned the mechanical stresses in the Earth's crust would have been released, allowing the surface to crack and split into various plates. But this would not have occurred until just after the age of the dinosaurs. The first part to crack off was a large mass which drifted southwards to form Antarctica (fossils of dinosaurs have been found in Antarctica).

Another plate separated to form Australia. A small mass broke off from the Australian plate and drifted northwards to form the Indian subcontinent. In the west, a large piece broke off from Africa to form South America (although not until early primates had evolved, as well as cats and snakes). Slightly later another part broke off from the northwest edge to form North America carrying with it more developed mammalian species. Thus the mechanism of plate tectonics developed as the long term end result of that giant cataclysm. The initial movements of the plates were much faster than their present rate.

The Earth is still struggling to regain perfect balance, as it is this which is responsible for continental drift and the continuing disruption as the various plates collide and subduct.

It is possible to calculate a crude estimate as to the size of the asteroid which struck the Earth and unleashed the cataclysm. The calculation is based on three assumptions: that the meteorite was travelling at almost 4 times the Earth's orbital velocity, that its density was 5 tons per cubic metre (much the same as the average for the Earth), and that half of the energy released was used in expelling the mass that was to become the moon. The present escape velocity from Earth is 11.2 km per second. But then there was no oxygen in the atmosphere. Atmospheric pressure was 21 per cent lower than it is now, and the frictional resistance was also 21 per cent lower than now. To allow for this, the escape velocity is 10 per cent lower. This may be a significant underestimate. The mass of the moon is 7.35×10^{22} kg. For this mass to achieve escape velocity requires 4.63×10^{30} joules of energy.

Standard laws of physics make it possible to calculate the mass of the asteroid from knowing the impact velocity and the energy released. With this and the density figure, it is possible to calculate the volume and so the radius of the asteroid (if it was spherical). With all these assumptions, a crude estimate of the size of the asteroid is that it had a radius of 203 miles. That is, if the asteroid was spherical its diameter was less than the size of Great Britain. If Earth's velocity contributed to the impact energy, as in a head on collision the asteroid would have been even smaller.

The Earth's Atmospheric Oxygen

Although the Earth's mass consists of 50 per cent oxygen, the oxygen is locked in water, carbonates, and oxides of various kinds, as well as silicates and phosphates. When the Earth emerged as a stable planet there was little to no free oxygen in the atmosphere. Evidence from the oldest iron ore deposits shows that what iron was combined with oxygen was combined with the smallest amount of oxygen possible. Now 21 per cent of the atmosphere is free oxygen. In the atmosphere there is 2.25 tons of oxygen over every square metre of the Earth's surface, totalling 10^{15} tons.

A popular explanation is that this arose from photosynthesis of micro plankton in the sea. This cannot be true. It flouts one of the basic principles of evolution's natural selection process, and there is simply no evidence that this happened (and there should be at least in the most ancient of rocks that contain early fossils).

Natural selection leads to the evolution of the most energy efficient life forms. Photosynthesis developed to form nutrients which could be stored for later use. In photosynthesis, oxygen is released, but in the latter usage of those nutrients, as well as at night, the oxygen is used for metabolism and growth, and the nutrients are converted back to carbon dioxide. There is a little leeway if some of the synthesised products are used for structural purposes, cellulose and lignin, but compared with the lifetime turnover of oxygen, this amounts to a very small fraction. When organisms die, they usually decompose back to carbon dioxide, unless the dead organism is buried in conditions which prevent this. One such mechanism is the formation of coal. But coal comes from dead trees, which grew when there was plenty of oxygen in the atmosphere. The life forms that ended up as oil could have contributed to atmospheric oxygen, but the worldwide deposits of oil or oil shales simply cannot account for the 10^{15} tons of oxygen in the atmosphere.

However, the real problem with this theory is the absence of raw carbon. The theory relies on the conversion of carbon dioxide to oxygen. This would have required that the atmosphere started with 21 per cent carbon dioxide. Even allowing for the increased solubility of carbon dioxide in water, the air/water partition coefficient means that the atmosphere would have had at least 2.5 per cent carbon dioxide as a gas. This is such a potent greenhouse gas that when taken with the enormous heat in the Earth, the oceans would have boiled after getting rid of its Snowball Earth characteristics. Earth would have become like Venus. Furthermore, in the oldest rocks that contain fossils, such as found on the east coast of Newfoundland and at the edges of the Laurentian Shield, for every 32 tons of oxygen produced there should be more than 12 tons of raw or un-combined carbon. In square metre terms this amounts to between 0.6 and 1.2 tons of free carbon per square metre (the range depends upon how much of the Earth's surface was covered by water for the proposed plankton to survive). These rocks contain very primitive fossils of multicellular organisms. Such organisms require a minimum of 15 per cent oxygen to survive. They were also sea living creatures and would have lived off the plankton. Yet the rocks they are found in do not contain anything like the amount of free carbon that the photosynthesis theory demands.

This leaves the solution of the problem as to the source of the world's atmospheric oxygen being outgassing from within the interior of the Earth. Carbon dioxide is a common gas which is released in large quantities by volcanoes. The chemical properties of oxygen preclude the accumulation of large quantities of free subterranean oxygen during the formation of the planets. But there is one mechanism which could do this. At 1000° Celsius, sand, silicon dioxide, dissociates into silicon and oxygen. This is used widely in the manufacture of computer chips. With the events that led to the formation of the moon, large quantities of sand or sandstone would have fallen into the magma to be heated. Each cubic metre of sand contains just over two and half tons of oxygen. A cubic kilometre of sand would contain 2.5 billion tons of oxygen. The world's atmosphere contains 2^{15} tons of oxygen. Oxygen would have been released as sand and sandstone fell into that big crater, but then some of the oxygen would have been trapped by the solidifying cooling magma and, as the crust thickened, would have remained trapped.

About a billion years ago enough heat had generated to melt the ice covering the Earth and, more importantly, melt the under-surface of the thickened crust. As a consequence volcanoes could now form. They did so on a massive scale. The Deccan traps as well as similar features in northeast Siberia show this. This would have allowed the degassing and releasing of the oxygen. This most likely started about 1 billion to 1.5 billion years ago. At the same time primitive anaerobic unicellular life forms began to evolve from the primeval soup that had been created by those early massive storms and then frozen in Snowball Earth. The oxygen would have killed them; 1 to 2 per cent oxygen is a deadly poison to anaerobes. But as the oxygen level slowly rose, some anaerobes evolved an adaptation to develop a resistance to the oxygen (much as bacteria evolve a resistance to antibiotics). Some went further to develop biochemical systems which could use this oxygen. This gave such organisms a very considerable evolutionary advantage. It allowed evolutionary experimentation into multicellular forms with cellular differentiation; 650 million years ago the oxygen concentration had risen to over 15 per cent, enough to enable mobile multicellular life to form. This was the Cambrian explosion of life and led to the greatest evolutionary experimentation into radically different structural organisations of life forms the world has ever seen.

Thus the starting point for the presence of life on Earth was that asteroid impact that led to the eventually release of oxygen from its combination with other elements such as hydrogen or silicon.

The Earth's Heat

Heat radiation measurements show that Earth is producing considerably more heat than it receives from the sun. Measurements of the Earth's heat production have been reported as being twice that forecast from the decay of the Earth's radioactive contents. If radiant heat is 1 to 2 per cent of the energy released by time slowing (the rest being mainly gravitational energy), the time slowing is producing between 11,250 and 22,000 kge of heat energy per second(i.e. 1/333,000 of that produced by the sun where 333,000 is the relative extent of the mass of the sun compared with the Earth). That is, the expansion of time through its effect on mass could account for the difference between what has been observed and that calculated from radioactive decay.

Chapter Six

The Moon and the Tides

The gravitational energy of the moon is responsible for the ocean's tides. The expansion of time allows a quantitative determination of the amount of gravitational energy released per second from the moon. From this, the amount of that energy intercepted by the Earth can be calculated. Gravitational energy interacts with mass. The greater the mass, the greater the interaction; that is, gravitational energy is used to accelerate that mass. As a result the acceleration produced by gravity is independent of the gravitational force. For a body with a mixture of masses with different relative densities the gravitational energy will be distributed in proportion to the absolute mass of each of the components. Thus the oceans, which make up only 0.022 per cent of the Earth's mass, will receive 0.022 per cent of the moon's gravitational energy that has been intercepted by the Earth. Over a twelve-hour period that energy will be used for two tides, one on either side of the globe. (The thirteen-hour intertidal period is a distraction caused by the moon's orbital rotation around the Earth. That is, relative to the moon, the high tide point has had to chase around the globe because its Earth-based reference point has apparently moved.)

Because of the spherical shape of the globe, that gravitational energy will distributed as though it was a dome overlying a ball. The energy will raise a dome of water over the ocean. Because of the rotation of the Earth, the apex of that dome will move across the ocean in a westerly direction, but the height of that apex will be unchanged. The apex corresponds to the point of maximum gravitational energy input per square metre to the ocean. The volume of the dome, in energy terms, will be, for each tide, half of the moon's gravitational energy that is distributed to the oceans over the twelve-hour period. The base of the dome will be half the surface area of the Earth. From then on, calculating the height, in energy terms, of the apex of the dome is standard geometry.

As a result the energy over the point of high tide is known. That energy will raise a column of water whose height is directly related to the energy supply and the relative density of the sea. The height represents the potential energy that would be released and the water column collapsed if the source of the energy input was withdrawn. The mathematical details are given as a mathematical model in the appendix to this chapter. The

Chapter Six

figures forecast the expected change in height between low and high tide in a big ocean such as the Pacific Ocean when the moon is more or less directly overhead—such as close to the Equator. The mathematical model ignores the slight oblateness of the globe, as this would have very little influence on the final result.

The model itself relies on seven independent key numbers. The first three are the value of the Earth's gravitational force, the relative density of the sea, and the average distance between the Earth and the moon. These three have all been measured. The next two are the estimated mass of the Earth and the estimated mass of the moon, both of which have a very good provenance. The sixth figure, the volume of the oceans, is more problematical, but surprisingly, this is a very robust number that errors of up to 10 per cent affect the result by only a few centimetres. This leaves the last figure, the rate of change of time per second due to the expansion of time. If the model fails to bring an answer reasonably close to the true answer, then the whole theory of the expansion of time fails.

The model forecasts that the expected difference between low and high tide for islands in the Pacific Ocean close to the Equator is 1.38 metres.

In reality the height of the tides are influenced by several factors, chief of which is the local geography. An extreme example of this occurs in the Bay of Fundy. If the measuring point is at the head of a narrowing inlet, the height of high tide will be exaggerated. There are also seasonal factors: spring tides, neap tides, and so on. Five islands in the Pacific Ocean near the Equator were randomly selected. Each island's weekly tide table was obtained from the Internet. From each table, a pair of high and low tides was selected, again randomly. The results are given in Table 6.1. Statistically if the differences are comparatively small and are occurring at random then the mean should be close to the true mean.

Table 6.1. High and Low Tides from Five Pacific Ocean Islands

Island group	Low tide	High tide	Difference	
Galapagos	0.50	1.9	14	Metres

Chapter Six 83

Christmas Island	0.21	1.26	1.05	Metres
Gilbert Islands Kiribati	0.24	1.97	1.73	Metres
Cocos Island	0.30	1.2	0.90	Metres
Marshall Islands Arno	-0.13	1.4	1.53	Metres
Mean difference			1.32	Metres

The result shows that the model predicted that the mean high tide would be 1.38 metres compared with the 1.32 metres, which is within 4.5 per cent of the predicted value. For a small sample this is ample validification of the expansion of time hypothesis as well as the calculation of the age of the universe in cosmological time.

The Background Microwave Radiation

The Earth is surrounded by microwaves that come from all directions. The microwaves cover an extremely small range of very slow frequencies. That is, they indicate an extremely low temperature, and the temperature range is around 1°. Their source must be extremely cold. This immediately raises the question, "What is glowing?" even if it is glowing very dimly. It has been suggested that the microwaves originated from a very hot temperature, such as the Big Bang, and that the microwaves have undergone some form of stretching. This hypothesis is untenable. Electromagnetic waves, once released, cannot change their frequency, which in turn is relative to the pace of time in the area through which they are passing. The Doppler or red shift, and the colour spectra of very distant galaxies, show that this must be true. A train of microwaves, if they became stretched, means that the leading waves are travelling faster than the hindmost waves. That is, the speed of light of a photon in transit can vary. This contradicts the very foundation of all relativity theory. But the successes of relativity theory are well shown in all nuclear reactors. The speed of light is fixed. Microwaves cannot change their frequency once in transit. They can be stretched whilst they are being emitted—producing the famous red shift. They can also be shortened if they pass from a zone where the pace of time is faster than it is on Earth, or its equivalent if they are passing through a medium where the velocity of light is slowed, as in clear water.

Chapter Six 84

Another aspect of microwave radiation is that the amplitude of the microwaves is the same despite coming from all directions. Since amplitude depends upon the energy of the original temperature and the distance travelled by virtue of the inverse square law, it means that the waves seen in the observed microwave radiation have all travelled the same distance, whether coming from the east, west, north, or south, from above or below. The constancy of the amplitude means that the emitting source (or sources) is more than 10 light hours distance. This is because travelling across the diameter of the Earth's orbit (almost 17 light minutes) makes small difference to the amplitude of the microwave. Thus a source that was at a distance of 15 light hours means that, according to the inverse square law, the additional time spent crossing the Earth's orbital diameter would reduce the amplitude of the microwave by approximately 4 per cent.

On the other hand, the microwave distribution shows certain areas where the temperature is slightly raised (about up to 1°) above the normal microwave temperature. These so-called hot spots can have an angular size almost that of the moon. The moon has an equatorial diameter of a little over 0.01 light second, and its distance is just over 1 light second. Thus if the hot spot was 100 light years away, just outside our galaxy, its diameter would be 1 light year. Such a large volume, if emitting energy, must contain some form of mass if it is static, and the gravitational effects should be noticeable. Even if the hot spot was 1 light year away, it would have a diameter of 0.01 light year. Clearly the hot spots must be much closer to Earth. Their angular diameter places the hot spots at a distance of, or just beyond, the distance of the Oort clouds.

The only conclusion must be that the solar system is surrounded by a shell of dust at or about the distance of the Oort clouds, that dust having a very low density. The microwaves are the result of universal time slowing, and the extremely low temperature is due to the very low density of the dust. The only reason we can measure it is because some of the "heat" radiation from the dust goes inwards, focussing on the centre of the solar system, and the detectors are receiving what is in effect an amplified signal. The hot spots are regions where there is a greater density of dust. Final proof of this suggestion would be it could be shown that the coordinates of some of the hot spots coincided with some of the Type 1A supernovas and that the magnitudes of their light was slightly greater (dimmer) than expected for their respective velocities. That is, for this type of supernova, all such supernovas should have the same magnitude when at the same distance, if they are to be standard candles.

The solar system arose from the explosion of a supernova. Such explosions generate many atoms of various masses, but the lightest weight ones are thrown, by radiation pressure, the farthest distance. They would have continued travelling but were caught by the gravitational pull of the remnants of the supernova. This created a shell of dust whose thickness varies randomly. Time expansion through its effect on mass would release small amounts of heat energy from the dust.

The alternative explanation, that the microwave radiation is a relic of the so-called Big Bang, does not bear critical examination. That theory requires that waves originated at one spot which was at a very high temperature and that this was at a period of time when the universe was extremely small, so that the microwaves would achieve uniform mixing if they were being reflected by some form of coating forming the boundary of the nascent universe. Furthermore, the microwaves initially were travelling at above the speed of light (inflation?). The hot spots were due to random parts having a slightly higher temperature than the rest of the mixture, and they have stayed that way ever since. Graham's law of diffusion requires that particles, which must include photons, diffuse according to the partial pressure or concentration of those particles. That is, the density of the heat energy of the source would fall faster the higher that initial temperature was. Moreover energy does not stand still. It expands and travels at the velocity of light. Amplitude (energy density) will change but not frequency. To obtain uniform density and so uniform amplitude, the distances travelled must be the same. It follows therefore that radiation coming from the east should not have the same amplitude as radiation coming from the west. Thus the hypothesis that the Big Bang was the source of the background microwave energy breaks well-proven laws of physics. Another name for this is invoking magic. This same description describes the so-called inflationary period used by some cosmologists to describe the very early universe.

APPENDIX 6

A. The Change in Land Level after Asteroid Impact

	Radius (km)	Volume (km³)	Mass (kg)
Moon	1,738	2.20×10^{10}	7.35×10^{22}
Earth	6,378	1.09×10^{12}	5.98×10^{24}
Total		1.109×10^{12}	6.05×10^{24}

Original Earth radius 6,421 km

The mean change in the Earth's radius means that the land level fell by 43 km.

B. The Size and Depth of the Crater

Atlantic Ocean area 80.6×10^6 square kilometres
Pacific Ocean area 163.8×10^6 square kilometres
Total area 244×10^6 square kilometres
Volume of ejectate 4.40×10^{10} cubic kilometres
Depth of cone 540 km

The depth of the crater is based on the assumption that a volume of mass equal to that of the moon was ejected from the crater but failed to reach the Earth's escape velocity and fell back to Earth some distance from the crater.

C. Energy of the Impact

Escape velocity (assumed)   10,080 m/sec
Energy for mass of the moon required to achieve escape velocity from e $= mv^2/2$
Energy purely for the moon $= 7.35 \times 10^{22} \times 10,080^2/2 = 4.15 \times 10^{30}$ joules
Total energy (assume 50% is used for ejecting the moon) 8.30×10^{30} joules

Chapter Six

D. Mass and Size of the Asteroid

The calculations assume that the asteroid has about the same density as the Earth and that the asteroid was spherical in shape. It also assumes that striking the Earth made no difference to the Earth's orbital velocity, as relative to the Earth its mass was far too small. If some of the impact energy arose from the Earth, this was too insignificant to materially affect the calculations, although the effect would be to reduce the mass and size of the asteroid slightly.

Earth's orbital velocity 29.77 km/sec
Impact velocity 148,850 m/sec
Total impact energy 0.5 × mass × impact velocity squared
Asteroid mass 8.32×10^{17} tons
Relative density (assume same as Earth) 5 tons per cubic metre
Volume of asteroid 1.66×10^{17} cubic metres
Radius if spherical 325 km (or 203 miles)

E. Calculating the Peak High Tide in the Pacific Ocean

The following algorithm is based on the assumption that time is expanding and causing mass to be converted to energy, 99 per cent of which is gravitational energy. That energy is radiated out in all directions, and so it should therefore be possible to determine the amount of the moon's gravitational energy intercepted by Earth.

Because the Earth is a mixture of masses, core, magma, crust, water, that gravitational energy is distributed according to the relative mass of the various components. Owing to the Earth's curvature, the gravitational energy sits as a domed cap, generating a dome of water on the ocean surface. The apex of the dome moves with the rotation of the Earth. The height of the dome in energy terms determines the height of the tide. This is calculated from the potential energy of raising a layer of sea water 1 mm thick by 1 square metre to a height of 1 mm. The number of layers gives the height of the tide.

Chapter Six 88

An algorithm to predict the peak height of the tides in the Pacific Ocean:

Stage 1. Determining quantitatively the moon's gravitational energy (The capital letters in the action column refer to the mathematical action to be taken with the item identified with the capital letter)

Item	Description	Action	Result	Units
A	Mass of the moon		7.47×10^{22}	kg
B	Age of the universe (cosmological time)		5.33×10^{18}	sec
C	Rate of loss of moon's mass A/B		14,006	kge/sec
D	Amount used for gravitational energy	$0.995 \times C$	13,936	kge/sec
E	Earth-moon distance (mean)		384,400	km
F	Area of sphere of radius Earth-moon distance	$4 \times \pi \times E^2$	1.86×10^{12}	sq. m
G	Radius of Earth		6,378	km
H	Cross section area of Earth	$\pi \times G^2$	1.277×10^8	sq. km
I	Earth as fraction of sphere area	H/F	6.88×10^{-5}	
J	Amount of energy intercepted by Earth	$I \times D$	0.96	kge/sec
K	Amount per 12 hours	$J \times 4,800 \times 9 \times 10^{16}$	3.7×10^{21}	joules
L	Surface area of Earth	$4 \times \pi \times G^2$	5.1×10^8	sq. km
M	Mean depth of oceans		2.5	km
N	Volume of oceans	$L \times M \times 10^9$	1.28×10^{18}	cubic m
O	Relative density of the sea		1.036	kg/m³
P	Mass of the oceans	$O \times N$	1.32×10^{21}	kg
Q	Mass of the Earth		5.976×10^{24}	kg
R	Oceans as a fraction of Earth's mass	P/Q	2.221×10^{-4}	
S	Amount of energy used by oceans per 12 hours	$R \times K$	8.25×10^{17}	joules
T	Amount energy per tide	S/2	4.13×10^{17}	joules

Stage 2. Calculating the peak height of the tidal energy which is distributed as a dome over half the surface area of the Earth. The equation for the peak height, h, of a regular dome with a circular base of radius r is Volume = $2 \times \pi \times r^2 \times h/3$

U	Base area of dome	$2 \times \pi \times G^2 \times 10^6$	2.55×10^{14}	sq. m

V	Volume of dome (energy × area)	U × height/3 = T		
W	Height (h as energy) of peak square meter	3 × T/U	4,848.7	joules

Stage 3. Calculate the potential energy required to lift a layer of sea water 1 mm thick by 1 square meter to a height of 1 mm against Earth's gravity (g). The number of layers is the sum of an arithmetic series 1-n. This can be rearranged as a quadratic equation $0 = n^2 + n - 2h/\text{potential energy}$. Calculating n yields the number of 1 mm layers, that is, the height of high tide in mid-Pacific Ocean.

X	Potential energy of only one layer	O × g × 0.001/1000	0.01015	joules
Y	Number of 1 mm layers	Solve quadratic equation	1,381	
Z	Peak high tide	Y/1,000	1.381	m

That is, the projected high tide of islands in the mid-Pacific Ocean is 1.381 metres. The figure relies crucially on the age of the universe in cosmological seconds, and so the rate of the expansion of time, leading to the quantification of the gravitational energy released from the moon and how much of that energy Earth intercepts. Observed results from a sample of islands (see above) was 1.32 metres

CHAPTER SEVEN

Jupiter and Saturn

Jupiter is a massive gas giant with a volume of 4.6 x 10^{14} cubic km, compared with Earth's 2.8 x 10^{11} cubic km. That is, it is 16422 bigger than the Earth but its mass is only 318 times that of the Earth. Its density is 1.25 times that of water. This rules out any large quantity of rocky material or even that the gases that make up Jupiter can be compressed to metallic form in any significant quantity. This raises the question as to why not? Jupiter is an emitter of radio waves, suggesting active electron activity within its core. It also has a magnetic field, which is strongest around its polar regions. This zone is also responsible for radio emission. However, the most significant functional feature is its very high rotation rate. At the equator this corresponds to a Jovian day of 9.9 hours. Towards the poles the rotation is slightly less (i.e., there is some differential rotation).

Jupiter's Equatorial Storms

At its equator, the circumference of Jupiter is approximately 450,000 (448,078) km. With a rotational period of 9.9 hours the circumferential velocity is 12.75 km/sec, that is, approximately one billionth of the speed of light. From the corrected Einstein's mass velocity equation, for every kilogram of gas entering from the polar region to reach the equator, the energy released is 80 million joules. At very low pressures, as at the surface of Jupiter, a kg of this gas has a volume of around 10 cubic metres; 98 per cent of energy released will be gravitational energy. Gas entering from latitudes nearer the equator will lose less mass, and so there will be proportionately less energy released per 10 cubic metres. But there are many millions of cubic metres being drawn towards the equator. That is, at the equator there is a considerable amount of gravitational energy being released. This will draw gas towards it, and that gas will also be subject to the effects of the high rotational velocity. Heat energy will also be released, causing the gas to expand and rise, raising the surface—and so

90

contributing to Jupiter's oblateness as well as creating an equatorial bulge. At the equator itself the gravitational energy, released as a consequence of the gas input from the lower latitudes, will pull mass from adjacent parts of the equator, which will add to the rotational velocity, causing the release of more gravitational energy and more heat. This will perpetuate right around the equator. The raised hot gas will be exposed to the temperature of space and cool and will diffuse towards the poles. In doing so the gas eventually enters a zone where the rotational velocity is appreciably less. This effect will increase further the drift from the equator. The gas will be drawn back towards Jupiter's main mass. As the cold gas drifts towards the poles, the reduced rotational velocity will cause the gas to absorb energy for the molecules of the gas to regain their original mass. Thus there will be substantial energy differential, both as heat and as gravitational energy between the tropical and subtropical zones. It is this which precipitates, fuels and perpetuates the massive storms seen in the equatorial zones. This effect will be seen both north and south of the equator, but the rotation of the storms will be in the opposite directions.

These storms consume an enormous amount of energy. Given the distance between Jupiter and the sun, solar energy is woefully insufficient for this. Equally, Jupiter's density precludes any radioactive source of energy. That is, the presence of these giant storms provides proof of Einstein's mass velocity equation and the inverse of his time velocity equation, so confirming the inverse relationship between time and mass. There is no other available source of the energy.

Jupiter's Heat

The temperature of the surface of Jupiter is variable but averages around 125°K. In contrast the surrounding space has a temperature of around 3°K. Any radiating surface which can maintain a temperature differential with its immediate surroundings of around 122° is emitting several kilowatts per square metre of radiating surface. Jupiter has a surface area of at least 6^{10} square metres (and probably up to 5 times that figure because of its uneven surface caused by storms all over its surface). In addition, its famous red spot, which has a diameter greater than that of Earth, is a heat vortex, exposing the red hot mass of Jupiter's interior. A huge amount of heat energy is being projected into space from the red spot.

In 1995 the Galileo Orbiter that was circling Jupiter released a parachuted probe that fell into the planet. The probe sent back data for just under an hour and reached a depth of 150 km before data transmission ceased. At that point it recorded that it was being subjected to a pressure of 330 pounds per square inch and that the temperature was 426°K. Whether the probe failed because of the high pressure or because the parachute or its ropes had melted is a matter of speculation. But the probe confirmed that Jupiter is producing huge amounts of heat.

Its density precludes any radioactive source for the heat. Its solar thermal energy is negligible. Energy production on this scale requires some form of mass-to-energy conversion. This leaves only one source of energy: the effect of universal time slowing. Simple calculation shows that this produces heat from mass at a rate of between 0.00035 and 0.0004 joules per kilogram per second. Given Jupiter's great mass, this is around 6×10^{23} joules per second or approximately 7 million kge as heat. But heat production accounts for only up to 2 per cent of the energy formed from mass by universal time slowing. That is, Jupiter is losing around 350 million kg of mass per second due to time slowing. However, its mass is periodically added to by the interception and crashing of comets and possible substantial quantities of interplanetary dust. This would be particularly so if the relatively nearby asteroid belt was the consequence of the disintegration of a nearby planet, as described in the previous chapter.

Jupiter's Moons

Jupiter has sixty-four moons, with the overwhelming majority being only a few kilometres in diameter. They appear to be a mixture of ice and rock. During the very early period of the solar system's history, the proto-planets swept through and accreted everything in and around their orbital paths. That is, most of Jupiter's moons were formed well after the formation of Jupiter. This fits well with the hypothesis that they are fragments of the planet that exploded and shattered and was responsible for the asteroid belt, as discussed in the previous chapter. Some of those fragments whose trajectories were away from the sun would be captured by the gravitational fields of both Jupiter and Saturn.

There is a very great difference between the masses of each of the four major or Galilean moons (Io, Europa, Ganymede, and Callisto) and the masses of all the other moons, with none of these other moons being of an intermediate mass. The Galilean moons are at least five orders of magnitude more massive than the largest of the sixty other moons. All but one of the non-Galilean moons orbit in reverse direction to the rotation of Jupiter. In contrast, the Galilean moons orbit in the same direction as Jupiter. The great difference in masses and their orbital motion strongly suggest that the Galilean moons had a different origin. There are significant differences in composition between the Galilean moons. But one thing they have in common is that they all produce large amounts of heat, and all have a molten interior of rock and iron.

Io, the nearest Galilean moon to Jupiter, has no water. It does however have many active volcanoes, more than 400. Io is 421,800 km from Jupiter. That is, the acceleration due to Jupiter's gravity is 0.7 metres per second at Io's distance. Since Io has a solid surface, the surface cannot be pulled from its substructure. For a solid object, the gravitational acceleration is uniform throughout the mass. There cannot be tidal flexing of the surface. In any case, tidal flexing requires that Io should be rotating very quickly. There is no evidence of this. This point has been examined in detail, as it has been suggested that tidal flexing is the source of Io's great heat.

In a previous chapter it was shown that the sun's gravitational energy is used to maintain the planets in orbit. If solar gravity suddenly ceased, the planets would continue in straight lines at a tangent to their orbital paths. This enabled calculation of the amount of solar gravity energy used to keep each of the planets in their orbits. The same can be applied to the Galilean moons as is shown in the table in the appendix to this chapter It will be seen that the amount of gravitational energy each of Galilean moons directly intercepts from Jupiter is less than the amount required to keep them in orbit around the planet. In this they are imitating what happens with Venus's orbit. Additional gravitational energy waves that would bypass the moons must be recruited, as each of the moons have absorbed all their intercepted gravitational energy. This would maintain an overall united front of all the gravitational energy radiating out of Jupiter. The important conclusion is that there is no evidence of any gravitational heating to the extent of melting rock.

Nevertheless, with all its volcanic activity, Io is very much hotter than the other three Galilean moons. All produce significant amounts of heat energy as a result of universal time slowing, but Io must have additional heat production. The most obvious source is that it must contain reasonable quantities of radioactive material whose heat is responsible for the volcanism. In this, it is just like the Earth. Its source must therefore be from the disintegration of a nearby rocky planet.

Jupiter and Its Comets

Jupiter has a long history of collecting comets, whether they are thrown up from the asteroid belt or inward falling from the Oort clouds. Most of these comets end by crashing into the planet, though some are captured and orbit around Jupiter in very highly elliptical orbits before crashing.

A particularly famous comet was Shoemaker-Levy, which crashed into Jupiter in 1994. In 1992, it showed signs of breaking up, but the fragments were comparatively close. When the collision came in 1994, it took six days for the more than twenty fragments to impact Jupiter, such was the extent of the separation. The approaching velocity was 60 km/sec. In achieving that velocity from zero velocity, when at its furthest point from Jupiter then, according to Einstein's mass velocity equation, it would have generated approximately 3600 megajoules of heat, that is if the original mass had been 100 kg. The comet originally contained a lot of water as ice, and as the fragment separated, the jet effect of the release of high pressure steam from each of the fragments caused them to separate. This effect continued even as the comet was leaving Jupiter for its final loop, the interior of the comet was so hot. As it moved further from Jupiter, so it slowed and its mass was partially restored by absorbing energy, only for the whole cycle to be repeated. What is extraordinary is that the fragments stayed almost in line throughout their two-year journey, although the impact sites on Jupiter were scattered across the southern half of the planet's face.

But in addition, universal time slowing would have added another million joules per year.

Gravitational Heating

For want of a better way to explain both Io's volcanism and the ignition of the stars, it was proposed that gravity induces heating sufficient to reach the hydrogen fusion temperature of the stars. Heating is the random movement or vibration of atoms and associated changes in the orbiting electrons of those atoms as a result of an input of electromagnetic energy. It is quite distinct from gravitational energy. Gravity induces motion in a particular direction, and the acceleration (and so motion) is irrespective of the mass of the object. That is, gravity can cause a rocky moon or planet to move in a particular direction, and if the gravitational force increases, the body will accelerate faster. But it is the whole body and not the individual molecules that move. If the body is solid, there can be no heating except under the special circumstances of very fast movement. Then relativistic time slowing occurs and with it, loss of mass, some of which will be as heat. This occurred with the comet Shoemaker-Levy. Friction has been proposed resulting from the movement. Friction requires contact between two surfaces which have different velocities. But gravity ensures that all the body, including all its component parts, move as one. Flexing will cause differential velocities within the mass being flexed, but gravitational flexing is very slow and ponderous and cannot generate enough differences in velocity between the various layers being flexed to account for the heat found in a body such as Io.

For liquids, viscosity can cause a differential velocity within a body of liquid, but again the amount of heat generated is trivial. Thus our oceans have tides and currents but the amount of heat that they generate is extremely small and rapidly dissipates.

With gases the gas laws come into play. Gravity can cause compression of gases, and this concentrates the heat energy into a smaller volume. This raises the temperature, but then the gas expands with temperature so that a volume of gas subject to compression will not only have the pressure being exerted upon it but it will also generate thermal pressure to oppose the applied pressure. Variable stars, which change their volume and brightness as a result of changes in their internal heat, are effective demonstrations that gravitational pressure can be overcome. At a more mundane level the equations of the gas laws (Boyle's law and Charles's law) show that compression will raise the temperature (see this chapter's appendix). Many gases subject to high pressure and having the heat removed condense into liquids, which are non-compressible. The condensation temperature of hydrogen and helium preclude this. With giant gas clouds of hydrogen, the gravitational pressure at the centre will be enormous, but as the temperature rises so thermal conductivity increases, dissipating the heat from the core to the outer parts of the giant gas cloud, so hindering the core hydrogen from reaching its fusion temperature.

Gravity describes acceleration. Acceleration of mass produces force. Force per unit area describes pressure. Thus gravitational force can induce a pressure which in turn will reduce the volume of a gas and so raise its temperature. But it requires very special conditions which interact with each other. The probability of all the many billions of sun-like stars having only small variations of those starting conditions to make gravitational heating as the sole cause by concentrating the initial heat energy is too small to be realistic. Thus gravitational heating of gas in giant gas clouds is not a realistic proposition for explaining the high temperatures within our stars. A different source of heat is necessary. The effect of time expansion on mass with the consequent release of heat provides a much better explanation for why stars are so hot.

Saturn

Saturn is the second largest gas giant, but its mean density is 0.8 times that of water. Nevertheless it has some metal at its core which is able to generate a magnetic field. As a result its south pole (which faces very slightly towards the sun) attracts sufficient solar particles to form gigantic auroras, which curiously can only be seen in ultraviolet light. Its orbital period is 29.46 years, and its distance from the sun is almost exactly 10 times the Earth-sun's distance. That is, its solar constant is 100 times less than the Earth's, amounting to only 14 joules per square metre per second. Light on the planet must be not much more than above a bright moonlight.

Chapter Seven

Its orbital velocity is 9.66 km/sec. To remain in orbit the planet has to be nudged towards the sun by a distance of 3.3×10^{-5} metres per second, as its orbital path is so long. As a result and because of its huge size and relatively low density, although being a very great distance from the sun the amount of gravitational energy from the sun that is intercepted by the planet is more than enough to keep it in orbit.

Saturn has an average temperature of 80°K in a spatial environment which has a temperature of around 3°K. Any radiator which can maintain a 77° temperature difference from its surroundings is putting out several kilowatts of energy per square metre. Saturn has an approximate surface area of 4.6×10^{16} square metres. In addition its relatively low density means that there is considerable amount of energy opposing its gravitational collapse. That can only come from heat, which must be substantial. Solar heat energy is trivial, being 0.2 per cent of 1 kge per second (assuming that none is reflected back into space). Saturn's low density precludes any significant amount of radioactive compounds, leaving the heat from time slowing being the only source of its heat energy production. Time slowing on such a mass as Saturn releases 2 million kge of heat energy per second. In total Saturn loses 100 million kg of mass every second to maintain its gravitational and heat energy production. Given that its mass is now approximately 5.7×10^{26} kg, it can afford that loss.

Saturn has a spectacular ring system in which are buried a number of small moons. The rings are made up of blocks of ice, and the gravitational effects of the small moons keep the rings stable to the extent they are often called shepherd moons. Saturn has three moons of particular interest: Japetus, Titan, and Enceladus.

Japetus: Two-faced Japetus is one of the most interesting moons in the solar system. It has two faces: the one which always faces Saturn is white, and the other is very dark, being a mixture of very dark red, dark brown, and black. The dark surface has occasional small patches of white ice, indicating that the dark material is shallow and probably only 10 to 20 cm thick.

Japetus is heavily pockmarked with craters which on the dark face are still covered with the near black material. Some of the craters on the dark side are several hundred kilometres in diameter. It has a large bulge going around its equator which consists of mountains that can be as tall as 20 km, making them the tallest mountains in the whole solar system. Its orbit is tilted approximately 15 degrees from Saturn's equator, and thus it is the only Saturn moon that can see all of Saturn's rings. It is the eleventh largest moon in the solar system, yet its mass is slightly less than one tenth of Earth's moon. Its density is fractionally over 1; that is, it is largely composed of ice with a small amount of rocky material.

The central bulge suggests that such high mountains have arisen from very strong compressive forces, much like the Himalayas on Earth. For the bulge to extend all the way around the equator suggests that the most likely cause is that Japetus arose from the slow collision of two roughly equal sized smaller moons, and the central bulge was consequential of the two centres of gravity moving to merge into one, compressing each of the contributors against each other.

There is a striking difference in temperature between the two surfaces. The dark faced side has a surface temperature of 130°K, whilst the bright face has a temperature of 100°K. Both temperatures are higher than Saturn's surface temperature of 80°K. There has been and still is considerable speculation as to both the origin of the dark material and why there is such a temperature difference. Clearly the heat must arise from within Japetus. Time slowing would ensure that the centre is very hot, enough to melt any rocky material. If Japetus formed from a collision, that would have generated even more heat. The possibility is that the rocky material contains radioactive material which, because of its density and centrifugal forces caused by its rotation around Saturn, pushed the dense radioactive material towards the distal face of Japetus, much like a very slow moving centrifuge. Thus heat generated from the decay of this material would warm that surface preferentially. It is strongly emphasised that this is yet another speculation about the origin of that temperature difference. It cannot be from heat from the sun, as the solar heat constant is only 14 joules per square metre and the surface temperatures from either face shows that it must be radiating considerably more heat than it receives. Similarly it cannot receive heat from Saturn, as the planet's surface temperature is colder.

The origin of the dark material is also a matter of speculation. Analysis shows that it is composed mainly of carbonaceous material, tarry substances much like the lakes on Titan. Sublimation of ice leaving a dirty lag has been suggested, the sublimation being caused by the heat from the sun. This cannot be true, since both faces over time have been exposed equally to the sun. The very large craters seen on the dark side suggest that there was substantial bombardment by meteorites and dirty comets. One possibility is that very late in the bombardment, Japetus was hit by several very dirty "snowballs," that is, icy comets with a high content of hydrocarbons. The heat of the impact would have melted and perhaps evaporated the snowballs only for the heavier hydrocarbons to condense and fall back on to the surface. A simpler and probably more realistic theory is that the dark surface is from debris from the small irregular moon Phoebe, which orbits in the opposite direction reasonably close to Japetus but is further from Saturn than is Japetus. Phoebe is heavily cratered and does not have much gravitational strength. Its irregular shape could be in part because it has lost a substantial amount of its mass from meteor collisions. Massive chunks could have been thrown up and, attracted by Saturn's gravity, began moving towards the planet, only for some of the big fragments to hit Japetus whilst en route. Japetus, always with one face always facing Saturn, would be hit on its distal surface (distal to Saturn). This could account for the very large craters seen on the dark side of Japetus. Subsequently, small bits of debris, thrown up either from the big collision which launched the major fragments or from later small collisions that formed additional craters, would easily escape from Phoebe and shower on Japetus, thus coating the distal face of Japetus and giving it its two-faced look. As neither the composition of the big meteorite nor the composition of Phoebe are known, this hypothesis can only be speculative, as indeed are all the other proposals.

Titan: Titan is the second largest moon in the solar system, with a radius of 2,575 km, only a fraction smaller than Jupiter's moon Ganymede (radius 2,634 km). For comparison, Earth's moon has a radius of 1,738 km. Titan's mass is 1.35×10^{23} kg; that is, it is 1.8 times more massive than our moon. It is 1.888 times more dense than water. That is, Titan has a significant rocky content. Its distance from Saturn is 1,222,000 km; that is, it is 3.2 times further from its parent planet than our moon is to Earth. It intercepts more than enough gravitational energy to keep it in orbit around Saturn.

Titan is unusual in that it is the only moon to have a cloudy atmosphere. This atmosphere extends farther from its surface than does Earth's atmosphere from Earth. The atmosphere is composed mainly of nitrogen and some hydrocarbons (which are responsible for the clouds). The atmosphere is sufficiently dense to support a parachute landing of a space probe. Sunlight on Titan is very dim, not much more than a bright moonlit night on Earth. Its surface varies from hydrocarbon lakes to patches of ice, with channels showing flowing liquid and lots of rather smooth pebbles. The observation of liquid hydrocarbon lakes indicates that Titan is relatively warm and that it has its own heat source, but it is too cold for liquid water. The most likely heat source is a combination of radioactive decay at the core plus heat released by universal time slowing. In this, it is repeating what occurs on Earth.

Enceladus: Enceladus is the most spectacular moon in the whole solar system. It is little; its radius is only ~250 km. That is, its size is about the same size as England and comparable with the meteor that was responsible for the formation of the Pacific Ocean and that produced the ejectate that became our moon. It is almost completely spherical; that is, early in its life it was able to shape itself into a sphere before it became frozen as a solid lump of ice. Its density is about that of water, so there is negligible rock in its core. Its surface temperature is 80°K, surrounded by space with a temperature of less than 3°K. With a surface area that at a minimum is around 800 billion square metres, it is radiating out a lot of heat energy. Its mass is 1.1×10^{20} kg, so that at its surface the gravitational force is only 0.18 m/sec/sec. Its orbit is almost completely circular, with a radius of 238,000 km from the centre of Saturn. Saturn's gravitational force at that distance amounts to only 0.67 m/sec/sec, so that tidal friction is minimal if it occurs at all.

Enceladus's northern hemisphere shows a fairly large number of impact craters. Its southern hemisphere is quite different. There are a few impact craters, but the rest seems covered by a very fine snow. There are a number of fissures on that surface, but particularly near the south pole are a series of four parallel fissures several kilometres long. These have been called tiger stripes. From these fissures come six jets of huge ice crystal fountains. These cloudy spumes reach more than 100 km from the surface. Such fountains, if on Earth, would be more than 2 km tall because of Earth's gravity. If they occurred on the Tibetan plateau, they would be as tall as Mount Everest. The fountains are ice crystals with a small amount of salt content. The crystals individually are very small, and most slowly fall back to Enceladus as gentle snow. It has been estimated that about 10 per cent of the fountain's mass attain escape velocity and so contribute to Saturn's E ring, in which Enceladus is embedded. From the size of the plumes and their spume, it is clear that they contain many thousands of kilograms of ice that started as superheated high pressure steam which condensed and froze almost immediately as it entered the cold of space. The lips of the fissures are noticeably warmer than the surrounding area.

The crystals contain, in addition to up to 2 per cent salt (but usually much less), traces of ammonia, hydrocarbons, and surprisingly, free nitrogen. The nitrogen is believed to have come from the degradation of ammonia—but that requires a lot of heat. This has raised speculation that perhaps there could be life forms inside a subsurface warm lake. This is highly unlikely; biological cells normally require an environmental salt concentration of 0.9 per cent as well as potassium.

This raises the question as to the source of the heat. It cannot be from radioactive sources. Enceladus is too far from the sun for any significant solar heating. Gravitational friction is negligible. This leaves heat produced by time slowing. Previous calculations have shown that the effect of time slowing on mass is to produce 0.003 joules per kilogram per second. Enceladus's mass of 1.1×10^{20} kg would provide 3.6 kge per second. Simple modelling of the energy consumption, such as allowing 10 kw of radiant energy per square metre and the kinetic energy of hurling 100 km (30 cubic metres) of ice per second in the ice fountains, as well as postulating that the centre of Enceladus is a cavity of radius 25 km full of superheated steam, accounts for about 10 per cent of the projected heat production from time slowing. These figures give a measure of the scale of the heat energy but also indicate that heat production is undermining the thickness of the ice walls of Enceladus. High pressure must be building up inside, and the ice fountains are acting as very inefficient pressure safety valves. The likelihood is that as the walls are thinned out by subsurface heating and melting and then the melt boiling, more fountains could appear. The rate of mass loss would dramatically increase, with the mass spewed out adding to the E ring of Saturn's ring system. If the internal pressure collapses as too much mass is vented out, Enceladus will become a hollow sphere. Gravitational collapse would follow, and the shell would be broken into many fragments which would thicken the E ring even more. It is Enceladus's small size which prevents its internal temperature from becoming high enough to induce the dissociation of the water into oxygen and hydrogen and so eventually exploding.

It is suggested that this has happened with a number of the smaller ice moons and that this is responsible for Saturn's magnificent ring system. Due to time slowing, the insides of the ice moons are warming up until they boil. The high pressure steam is then vented off, with collapse and breakup of the hollow remnant.

APPENDIX 7

Gravitational Heating of Gas

Gravitational compression has been held responsible for the starting hydrogen fusion in great clouds of gas, thus generating the luminous stars we now see. It is possible to create a mathematical model to show that this is feasible, but it requires very limited and special conditions. If the temperature of the great gas cloud was 3°K (Kelvin), then to raise it to the presumed temperature of 1.5 million degrees required for hydrogen fusion would, according to Charles's gas law, cause it to expand by a factor of 500,000. Equally, applying pressure to compress a gas so that its heat energy is compressed in a small volume at a temperature of 1,500,000° would require raising its pressure by a factor of 500,000.

Charles's law gives rise to the following equations:

Equation Appendix 7.1 $V = v \times 3/273$ which on heating changes to
$V_1 = v \times 1,500,000/273$

Equation Appendix 7.2 $V_1/V = 1,500,000/3 = 500,000$

Boyle's gas law at constant temperature gives rise to
Equation Appendix 7.3 $V = v \times P/760 = v \times 1/N$ atmospheres of pressure

Inverting Equation Appendix 7.3 results in

Equation Appendix 7.4 $v/V_1 = v \times 1/500,000$ atmospheres of pressure

It follows that gravitational compression must overcome the effects of thermal expansion and then compress the gas by a factor of 500,000. Putting it all together:

Equation Appendix 7.5 $V = v \times (1{,}500{,}000/3)/273 \times 1/(500{,}000)^2$
atmospheres of pressure

If a gas cloud with a volume of 1 cubic light year started with a density of 1 mg per cubic metre and a starting temperature of 0.003°K, gravity could compress the cloud to reach hydrogen fusion pressure, but any temperature lower than this and the resulting pressure would be below that necessary. For a solar mass star, it would be necessary to start with a volume of gas with a radius of 350 light seconds and an average density of 10 grams per cubic metre and a starting temperature of 10°K before hydrogen fusion could occur. With the many billions of billions of G stars that exist in the universe, it is highly improbable that they all had similar starting conditions. There must be a controlling mechanism which defines how much gravitational energy is made available, as any compression would take a period of time, during which heat could escape by radiation.

The alternative is that it is not gravity that is uniquely responsible for the heat of the stars. There must be an additional source of energy which is pumping heat into the gas cloud. That source comes from the result of the effects of time expansion on the mass of the nucleons within those great gas clouds.

The Galilean Moons

The accompanying table gives data on the Galilean moons. It shows that none of the Galilean moons are able to intercept enough gravitational energy from Jupiter to remain in a stable orbit (although Europa and Ganymede intercept almost enough). That is, all the Galilean moons must recruit additional gravitational energy from the gravitons which would otherwise bypass them. The expansion of time releases more heat in Ganymede and Callisto than it does with Io, yet Io has volcanoes and the others do not. There is no available gravitational energy for gravitational heating. That is, Io must have an additional source of heat: radioactive decay.

Table The Galilean Moons

	Io	Europa	Ganymede	Callisto	Units
Mass	8.93×10^{22}	4.8×10^{22}	1.48×10^{23}	1.089×10^{22}	Kg
Orbital period	1.769	3.553	7.155	16.69	Days
Orbital period	1.53×10^5	3.07×10^5	6.18×10^5	1.44×10^6	Sec
Radius	1,821	1,560.8	2,631	2,410	Km
Mean distance from Jupiter	4.22×10^5	6.71×10^5	1.07×10^6	1.88×10^6	Km
Length of orbit	2.65×10^6	4.21×10^6	6.72×10^6	1.18×10^7	Km
Mean orbital velocity	17.32	13.73	10.87	8.20	Km/s
Nudge distance	0.36	0.14	0.06	0.02	Metre
Energy needed for nudge	6.28×10^4	5.27×10^3	2.51×10^3	1.91×10^2	Kge/s
Hypothetical shell area	2.23×10^{12}	5.65×10^{12}	1.44×10^{13}	4.45×10^{13}	Sq. km
Moon cross section area	1.04×10^7	7.65×10^6	2.17×10^7	1.82×10^7	Sq. km
Fraction of shell	4.68×10^{-6}	1.35×10^{-6}	1.51×10^{-6}	4.1×10^{-7}	
Gravitational energy intercepted	1.73×10^3	5.03×10^2	5.61×10^2	1.52×10^2	Kge/s
Heat generated by expansion of time	3969	2133	6527	4800	Kge/s

The hypothetical shell area is the surface area of a hypothetical sphere whose radius is the distance between Jupiter and the relative moon. In all cases the gravitational energy directly intercepted from Jupiter by the various moons is less than that needed to maintain the moon's orbit around the planet, thus Europa needs just over 10 times (5.27/5.03) × ($10^3/10^2$) the amount of directly intercepted gravitational energy from Jupiter simply to remain in orbit.

CHAPTER EIGHT

Chaos, Destruction, and Regeneration: A Revised Cosmology

Current cosmology is based on the Big Bang theory. In this theory the entire universe, its mass, its space, arose from a minute microdot, or singularity, of intense energy. Suddenly there was an explosion. Time started then. Space began to form, and the energy expanded faster than the speed of light. The energy then condensed into various subatomic particles which joined together to form protons and antiprotons as well as electrons and positrons. There was an imbalance so that the mutual annihilation left over some electrons and protons. The mutual annihilation released energy, which went through the same conversion process and mutual annihilation, leaving yet more protons, and so on until there were negligible amounts of antiprotons and positrons. Meanwhile, some electrons joined with some protons to form neutrons. Free neutrons have a half-life measured in minutes. They had to join with some of the protons, so forming helium nuclei. Some further merging took place so that a small percentage of the mass became lithium. The great clouds of mass continued expanding but now were moving at close to the speed of light. Gravitational attraction caused the great clouds to fragment and condense to form giant stars. These soon exploded as supernovas and formed a second generation of smaller stars, such as the common G stars. Meanwhile, gravitational heating caused the clouds of stars to ignite, so forming the bright stars which we now can see.

There are a number of serious difficulties with the Big Bang theory. One such is size. There are an estimated 10^{80} nucleons in the universe. The nucleon has a radius of $\sim 10^{-15}$ metres. 10^{80} nucleons, if closely packed together with no space, would occupy a volume with a radius more than 600 times bigger than our sun. Energy would have had to have expanded

Chapter Eight 107

to this size before protons could form. The justification for arguing that the expansion was faster than the speed of light is to account for the fact that on the broad scale, the distribution of mass throughout the universe is remarkably uniform. There is the same number of galaxies in view no matter which direction one looks at the universe. That is, when the protons were formed, they were remarkable evenly distributed.

But this question of moving faster than the speed of light across a radius of at least 250 light seconds is invoking magic. Magic may be defined as something which breaks the known and proven laws of physics. One such law is the speed of light. Calling this process by a fancy name, inflation, still does not take away the fact that what is being invoked is magic.

But there are other problems with the Big Bang. One is that it ignores the effects of gravity until very much later in the process. Gravity is one of the four (six, if one counts Burkhard Heim's acceleration force and gravito-magnetic force) fundamental forces of nature. As such, gravitational energy would be an important component of the energy contained within that singularity. The universe is estimated to contain 10^{80} nucleons. If these were all packed together, the resultant gravitational force would generate a velocity of 10^{20} metres per second at its surface. This is 10^{11} (or 100 billion) times faster than the velocity of light. Nucleons store gravitational energy and very slowly dribble it out over time. If protons had not yet been formed, then the gravitational energy available would have been 10^{53} times more. In fact the size of the universe would have to have a radius greater than 27 light years before light could be released.

An associated problem is the curvature of space. Einstein proposed that gravity causes a curvature of space which is minimal with masses the size of our sun, but with the very intense gravitational force at the beginning, the space would have been as in a very tight but long spiral. Anything travelling outwards would be caught in this spiral, and there effectively would be no expansion of the universe. The alternative is that Einstein's theory on gravity causing a curvature of space is wrong.

Another problem with the Big Bang theory is the question of time. It is posited that the universe is 13.7 billion years old. This figure is based on an extrapolation derived from the Hubble constant, which is assumed to be below 50 km/sec/Mpc. If the constant has a value of 52 km/sec/Mpc, the age figure comes down to 12.5 billion years. The Hubble constant shows that velocity increases with distance. The Big Bang theory implies that the energy behind the velocity of the receding galaxies was imparted at the Big Bang. Galaxies that were endowed with high energy (and so high velocity) would maintain that velocity as they moved outwards. Slower galaxies would maintain their lower velocities as they too moved outwards. That is, over time as the galaxies spread further and further outwards, the Hubble constant would become lower over time. Yet distant galaxies show the same Hubble constant value, as do comparatively near galaxies.

There are three more problems or paradoxes with the Big Bang theory. The first is the globular clusters. Within our own galaxies, there are over 150 distinct clusters of stars. Each cluster is composed of at least a thousand stars. For each cluster, the estimated age of the stars within the cluster is the same (although different clusters have slightly different ages). But the age range of the clusters is around 13 to 15 billion years. That is, some clusters are apparently older than the age of the universe.

The second paradox is that of distance. The Sloane star survey has shown that we are surrounded by galaxies that are more than 10 billion light years distant. Mathematically, the sum of the distances in light years of two galaxies that are on opposite sides of the universe cannot exceed the age of the universe. It would take time for the galaxies to reach their positions and then time for the light to travel back to Earth. The galaxies could not form their spiral patterns if the mass of the galaxy on its outwards journey was travelling at very close to the speed of light. It would take time, several billion years, for the galaxies to form their geometrical structures, then for stars to go through their life cycle before ending as supernovas that we can detect.

The third paradox is why it has taken so long for the light from these very distant galaxies to reach us. Going by the time table of the Big Bang, they were formed when the universe was very young and was, therefore, comparatively small; that is, they were comparatively nearby. It should not have taken 10 billion years for their light to reach us.

However, the final problem with the Big Bang theory is that it proposes that all nucleons were generated at the same time. As such, they should all have the same velocities if the force generating the velocity arose from the nucleons. This is incompatible with the spread of velocities of different galaxies. Their nucleons could not have been generated at the same time; the generation must have been spread over a very long period of time.

The Natural History of Black Holes

Black holes appear to come in three forms: (a) the standard black holes, which are the end result of a type 2 or giant supernova, (b) the supermassive black holes, which form the centres of spiral galaxies, and (c) the primordial black holes, which are very large but are collections of the six primary fundamental forces, and include one of which was responsible for the formation of our visible universe. That is, it is a polemical point whether there exist other primordial black holes producing their own universes well outside the range of our visible universe, or whether primordial black holes exist within our visible universe but are quiescent and held so by factors within the visible universe. We do not know (nor will we ever be able to know).

Black holes have two features in common. This is that they have an event horizon, whereby because of the enormous gravitational strength of the black hole, the force required for anything to escape results in a velocity that is greater than the velocity of light. But this is impossible. The only exceptions are gravitons, in that gravitational force is carried by gravitons, and gravitons do not interact with each other (although they align with each other forming continuous sheets of waves). The event horizon enables the determination of the radius and therefore the volume of the black hole. The other feature is that according to Einstein's general relativity theory, space is tightly curved around a black hole so that anything entering this zone will remain travelling in a circular orbit around the black hole. This zone of curvature extends outwards from the black hole, gradually lessening until, because of the inverse square law, the gravitational force produces an acceleration of less than 10 m/sec/sec. When this occurs, space becomes essentially flat or straight.

If a solar mass of hydrogen molecules were to collapse and form a black hole, it would have a radius of approximately 664 km (smaller than the size of England). Its volume would be 1.22×10^{18} cubic metres. From Avogadro's hypothesis of the number of molecules in a gram molecule of hydrogen (6.022×10^{26}), the number of atoms in a collapsed solar mass of hydrogen is 1.18×10^{57}, but these would be compressed into neutrons. Neutrons have a radius of approximately 1×10^{-14} metres. The total volume of neutrons of this solar mass of hydrogen, if compressed to a black hole, would be 5.57×10^{16} cubic metres. That is, each neutron would have approximately 20 times its own volume of empty space around it. It would have 1.2 times its own diameter of space before coming into contact with an adjacent neutron.

The bigger the mass of the black hole, the greater the separation of the neutrons. For a 4 million solar mass black hole at the centre of our own galaxy, each neutron has approximately 35 times its own diameter of empty space before coming into contact with another neutron. This distance is still considerably less than the distance of an atomic nucleus to its surrounding electrons. The mechanism underlying this separation is unknown, but the essential consequence is that the neutrons are independent of each other. This has a significant effect when considering the expansion of the visible universe due to time slowing. The Hubble acceleration force requires that each nucleon within a galaxy is accelerated at a rate of 5×10^{-10} m/sec/sec. That expansion is driven by the energy from each nucleon, just as gravitational energy arises from each nucleon. This in turn means that black holes within any galaxy, be they standard black holes or central supermassive black holes, keep up with the overall velocity of their containing galaxy as that galaxy takes part in the expansion of the visible universe.

More significant is that fact that the release of energy for that 5×10^{-10} m/sec/sec acceleration is dictated by the prevailing pace of time that is well outside the black hole. Time within a black hole must be normal, as must be time in the zone of curved space around the black hole. Time for a particle passing through or attempting to pass close to the black hole is slowed, just as it is when a particle is travelling at very high speeds. That is, for the nucleon the pace of its time arises from within the nucleon. This is a fundamental concept which seriously undermines the concept

Chapter Eight 111

of space-time continuum. Time and mass are inversely related. But mass is compacted energy. It follows that time is the inverse of energy. That is, time multiplied by energy would reduce the expression of that energy. Thus within the nucleon, time would inhibit the expression of gravitational energy, thereby controlling the release of stored gravitational energy.

Standard black holes are the result of supernova explosions of massive stars. If the residue of the supernova is greater than 8 solar masses, the result is a black hole. Time slowing causes mass to revert to energy. Within the black hole, with the exception of gravitational energy, that energy is trapped within the event horizon. As time passes, more of the mass reverts to energy, 98 to 99 per cent of which is gravitational energy. The remainder is heat energy, which becomes more and more pent up. The gravitons leak out but are trapped by the curvature of space around the black hole so that they orbit continuously close to the event horizon. There must come a point where the nucleons within the black hole cannot supply enough gravitons to maintain the event horizon, and the pent-up heat energy floods slowly (because of gravity) into the curved time zone to mix with the gravitons and condense to form new protons. The absorption of gravitons increases the flow of gravitons, quickening the demise of the black hole. The newly minted protons are held reasonably close together to form large clouds. Some of the protons eventually combine to form stars. This is the stellar nursery whereby new stars are "born."

Supermassive black holes have a slightly different origin. At the formation of a galaxy, as it condenses into a large cloud of neutrons, the nucleons will be pulled in the direction of any group which has the greatest gravitational force. But at the centre of the giant cloud, that pull is likely to be the same in all directions. The result is that at the centre, nucleons cluster together, forming a supermassive black hole. The gravitational strength is too great to form a giant star.

But elsewhere in the proto-galaxy, at random there will be similar scenarios, albeit on a much smaller scale, but forming massively huge giant stars, each containing the size of several thousand solar masses. Each giant star quickly "burns off" much of its hydrogen (that is, the hydrogen fuses in an attempt to keep the temperature down) so that it rapidly ends up as a giant, very helium rich star that will overheat and end up as a supernova. The resultant release of mass will form stars in close gravitational attendance—that is, the stars will form as a cluster of stars of approximately the same age, and there will be several thousand stars in the cluster. In different areas of the galaxy, there will be similar events, but the precise timing of the explosion of their supermassive star will depend somewhat on its size. The end result is that our galaxy will contain a substantial number of such clusters of stars, each one in the cluster showing the same age, although different clusters will have slightly different ages. In effect this is the creation of the globular clusters, of which over 150 exist in our own galaxy. Smaller but still large stars will echo this scenario, thus forming stellar nurseries.

Supermassive black holes are large. The 4 million solar mass supermassive black hole believed to lie at the centre of our own galaxy would have an event horizon of 4.42 light seconds. Its overall size is 8 times larger than our sun.

Primordial black holes are rather different. They do not contain mass but contain energy in all of its six forms that underlie the fundamental forces of nature: the weak and strong intranuclear forces, the electromagnetic force, gravity, the Hubble expansion force, and the Hubble magneto-gravity force. They are huge, with event horizons that run to many thousand light years. They are apparently timeless; that is, the immense gravitational force slows time to a standstill. The evidence for their existence is the observation that scattered throughout the universe are sites where the light from a more distant galaxy is duplicated, or even trebled or quadrupled, yet between the images is total blackness. The light directly from that distant galaxy does not penetrate, but rather the beams of light from that distant galaxy, that should have bypassed the object, have been refracted, producing multiple images of the same object. The dark gap between the images is due to some massive object which is obscuring the direct light yet has the gravitational strength to refract light beams which should otherwise bypass it. To obscure the direct light, the obscuring mechanism must be at least as big as the images that the distant galaxy would have at that distance. The only mechanism that could do this is a very massive black hole surrounded by its layers of curved space. This size is many orders of magnitude greater than that of conventional supermassive black holes.

Chapter Eight 113

At a gravitational acceleration of 10 m/sec/sec (i.e., Earth's gravity), there is effectively no curvature of light, whereas at the event horizon of a black hole, the gravitational acceleration is 3×10^8 m/sec/sec. The inverse square law means that the curvature horizon of a black hole is approximately 5,500 times the black hole's event horizon distance from the centre of the black hole. For our galaxy's central black hole, the curvature horizon is about 6.67 light hours from the centre. This means that the geometry of the visible universe is more than 99.99999 per cent flat: that is, it has no effective curvature of space apart from very close to massive bodies.

Black holes must also die. The smaller the black hole, the shorter its life. The reason is that black holes radiate out energy, gravitational energy. That energy must come from mass. The reduction in mass would cause the event horizon to shrink as the mass became smaller. Meanwhile, time slowing would be converting mass also to energy, such as electromagnetic energy as well as the energy of the weak and strong intranuclear forces. When the gravitational force becomes too weak to sustain the event horizon, the whole system would burst, releasing vast quantities of energy for a brief moment of time. These could the cause of the gamma bursters, that is, very sudden and very violent but brief bursts of gamma radiation that occur occasionally seen at high altitudes. The first panic reaction was that someone was testing an H bomb in near space, but detection of its possible location eliminated all possible Earth sources. The energy intensity of these bursters is very high, and that energy has obviously travelled huge distances. The only possible source of so much energy occurring so quickly is a very violent event that releases astronomical amounts of energy. The death of a black hole fits this description.

Key Factors in Cosmology

In any revised cosmology, there are a number of points which need to be taken into consideration.

a. That the age of the universe is 169 billion years in cosmological time, something that is confirmed when calculating the rate of loss of mass and so of gravitational energy production from the moon

Chapter Eight 114

        that predicted so accurately the height of the tides in mid-Pacific Ocean.

b. That time is expanding. Without this effect, the Hubble constant could not be constant.

c. That time and mass are inversely related. The resultant effect produces the heat inside the planets and accounts for the energy for the great icy fountains on Saturn's little moon, Enceladus. But mass is compressed energy which exists in quanta. That is, time must exist in quanta. The density of the time quanta must increase the pace of time. Any lowering of that density will slow the pace of time, creating the illusion that time is expanding.

d. That the life of any galaxy is around 25 to 26 billion years. This is because the Hubble acceleration constant, at 5×10^{-10} metres per second per second, eventually causes galaxies to approach close to the velocity of light. By then they will become faint dusty replicas of themselves, with negligible gravitational or light energy. There is no precise age because with the increasing velocity, as the velocity of the receding galaxies approaches the speed of light, for those galaxies time itself slows, perhaps prolonging its death throes by several billions of cosmological years.

e. The formation of stellar nurseries, both in the past that were responsible for the globular clusters, and now the more recently discovered stellar nurseries producing new stars.

f. There exist giant primordial black holes which obstruct the light from distant galaxies but also cause replica images of those distant galaxies.

g. Those primordial black holes were and are very big but effectively timeless. Some may still exist. They possess vast quantities of time quanta but the immense gravity renders them inoperative. Since these primordial black holes existed as imprisoned energy, there is no mass, making the Hubble expansion force inoperative. These

Chapter Eight

primordial black holes are stationary, in contrast to the black holes within galaxies.

h. That the distance in light years of galaxies on opposite sides of the universe cannot exceed the age of the universe in years. This is because it would take time for the mass of those galaxies to reach their current positions and then go through their life cycle before sending their light back to us.

i. Higgs boson. A boson, sometimes called the "God boson," is present in every nucleon and is responsible for binding the constituent particles of energy to a sufficient density to form mass.

Chapter Eight 116

Figure 8.1 The effect of a primordial black hole causing the refraction of the image of a distant galaxy, producing multiple images. The gap between the images reflects the size of the curvature horizon relative to the size of the image at that distance.

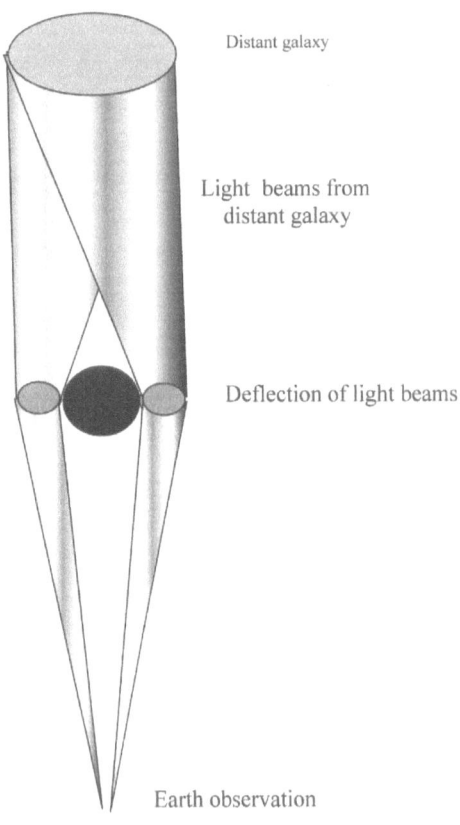

The deflection of light beams by a large gravitational object so producing duplicated or more images when observed with earth telescopes.

j. The variable age of the different galaxies. The most distant galaxies are travelling much faster than is our own galaxy or our nearby galaxies. It would take time for those far galaxies to build up speed. That is, the creation of galaxies must have been a long, drawn-out event.

Chapter Eight 117

Any revised cosmology must take into account all these essential factors.

A Modern Cosmology: The Evolutionary Universe

It is emphasised that the proposed modern cosmology represents an attempt to draw up a cosmology that takes into account these key factors. The following descriptions must therefore be regarded as only a reasonable alternative working hypothesis.

The proposal is one of chaos, nuclear generation, annihilation, and regeneration until order for coherent expansion evolved. This concept arises from questioning why the universe is expanding in such an orderly manner, with all galaxies moving outwards.

Certain general principles must also apply. Time must exist in quantum units. It follows that for balance there must be a quantum of time for every quantum of energy, irrespective of the kind of energy, be it electromagnetic energy or gravitational energy. The pace of time is dictated by the density of these time quanta which interact with all the forces, and all other particles and masses, from photons upwards, giving them their velocity. The effectiveness of these interactions is hindered by gravity, and the extent of this hindrance depends upon the local gravitational strength. This has an important consequence. If the pace of time depends upon the density of the time quanta, then every time the volume of the universe doubles the density is halved, and so the pace of time is halved. In addition every time the pace of time-halved mass is reduced by half, the mass is shed as energy. There is another factor concerning time quanta. High gravitational strength slows time by inhibiting the effectiveness of the time quanta. As a result the velocity of photons and therefore the velocity of light are slowed. Similarly, extreme cold will inhibit the effectiveness of the time quanta so that just as gravity slows the velocity of light, it has been shown experimentally that at a temperature that is a fraction over absolute zero, the velocity of light is reduced to walking pace. That is, it would take a photon over two years to travel what it would normally cover in 1 second. In effect this means that time, as we know it, would not start until space had been warmed up a fraction.

Chapter Eight 118

In the beginning was a primordial black hole. Its initial size is unknown, but eventually it had a radius that was half the radius of our own galaxy. This primordial black hole contained only the particles carrying the six fundamental forces of nature, including Hubble's expansion force and his gravito-magnetic force plus the time quanta. Because of the extremely high gravitational force, the effectiveness of these time quanta was negligible. The pace of time was therefore negligible but not absolutely zero. The velocity of the force carrying particles was also negligible. The particles carrying the other forces could not escape, partly because their velocities were so slow and partly because of the concentration of gravitons. But gravity cannot constrain gravitons (if it did, the spiral shape of our own galaxy would be impossible, as it would be for all similar galaxies).

Surrounding this black hole was space. Whether gravity caused this space to curve tightly or simply caused every and any wave particle entering this space to have its trajectory refracted so that it orbited the black hole is a matter of debate. The effect is the same. For convenience therefore the analysis assumes that space is curved by gravity.

The space was tightly wrapped and curved around the surface of the primordial black hole. The surface of the black hole was the event horizon, whilst much farther out was the curvature horizon, beyond which space was flat; that is, it had no significant curvature. The gap between the event horizon and the curvature horizon meant that any time quantum which entered this zone would be slowed. The thickness of the zone was approximately 5,500 times the radius of the primordial black hole's event horizon. Not only was space curved and wrapped around the black hole, it was also compressed. (If black holes can cause space to curve, then space must be regarded as a material which is capable of being stretched and so logically capable of being compressed. This means that space has a variable density.) There were no photons of electromagnetic energy; all that was confined to within the primordial black hole. As a consequence the temperature was either at absolute zero or a fraction above it. Because of this there was virtually no time in this tightly wrapped space. As one progressed outwards from the event horizon, so the inverse square law came into play, diminishing the force of gravity and also diminishing the rate of curvature of space.

Chapter Eight

Energy intensity gradients caused some of the very slow moving gravitons to drift to beyond the event horizon, but the very intense curvature of space caused them to very slowly but tightly orbit the black hole. The gravitons carried with them some of the time quanta, and it was this which gave them their ponderous velocity, but it was so cold that the gravitons hardly moved. But over what must have been a near infinitude of time, the black hole was enlarging to become a thick coat of gravitons, permeated only by the time quanta.

This had two effects. Within the black hole the mean intensity of gravity was slowly lessening, so enabling the time quanta very gradually to increase their effectiveness. This increased the velocity of all the force carrying particles, increasing the pressure to escape and forcing the black hole to expand. Outside of the black hole's event horizon, the accumulation of the orbiting gravitons had an overall effect of reducing the concentration of gravitational energy per unit area that was acting on space. This led to an infinitely slow but gradual lessening of the curvature of space. This had a positive feedback of increasing the thickness of the coat of gravitons orbiting around the black hole. The distant curvature horizon gradually moved nearer to the coat. But the overall density of the time quanta was extremely high, although the temperature was still close to absolute zero.

Eventually the expansion of this black hole meant that the event horizon could not be maintained, and energy carrying particles could travel into the narrowing space between the surface of the ex-black hole and the curvature horizon. This included electromagnetic carrying particles, key of which were heat photons. Now the time quanta could act. The high density of the time quanta accelerated the pace of time, although this was hindered by the strength of the gravitational field, but eventually time would become over a million times faster than it is now.

Some of these energy particles condensed into electrons and positrons. Their direction of movement was chaotic. Inevitably there were collisions between the electrons and positrons, reducing both back to their original energy components, which were scattered locally only to re-form. But the overall density of energy was sufficient that more electrons and positrons were formed, only to suffer for the same mutual annihilation effect. But there was a preponderance of electrons over the positrons (why this was so is not known), so gradually the positron population was reduced to insignificance. It was survival of the fittest; in this case, fittest meant preponderance of strength—just like two armies facing each other.

Since energy cannot be formed nor destroyed, energy from any annihilation was released into the space curved zone. Other energy particles started to come together to form subatomic particles, including Higgs bosons. The local strength of gravity was still strong enough to hold some curvature of space. The energy carrying particles previously trapped in the black hole were trapped in this curved space zone. The subatomic particles joined and formed various quarks, each one apparently being derived from its prime fundamental energy source. That is, there were (and are) six fundamental forces and six types of quarks. There were various combinations of quarks and other subatomic particles joined up, but almost all were unstable and so decayed into the background energy pool. This was the evolution of matter. Unless the subatomic particles could join up, they "died" or decayed back to naked energy. The energy would re-form back to the original subatomic particle, or if it was joined by other bits of energy, it could form a different subatomic particle. It was a matter of chance. A major factor in the chaos of formation and decay was that the life or duration of existence of any independent subatomic particle was very, very short. A variation which might prolong the survival of the subatomic particle was if it joined with a different subatomic particle and the combination had better survival prospects. Even so these were very slight. It is small wonder that it took billions of years before stable particles evolved, or rather collections of subatomic particles acting together could evolve as a stable bigger particle.

There were two particles, or rather specific collections of subatomic particles, which managed to do this. Protons and neutrons evolved as the Higgs bosons clamped the quarks together, forming mass. Even then the free neutrons were unstable and would disintegrate back to their primary energy components unless they joined with a proton within thirty minutes. It has been shown experimentally that free neutrons can only exist as free neutreons for up to half an hour. The protons had a charge which could be negative (antiprotons) or positive. When the two met there was annihilation, releasing substantial amounts of energy including gravitational energy and time quanta, facilitating the escape of yet more energy from the leaking, shrinking primordial black hole.

Energy cannot keep still. It would move in a random direction, following the curvature of space. Inevitably other patches of energy would collide, and together they would eventually evolve into protons or antiprotons. But like the electron/positron conflict, there was an excess of protons, and so the antiprotons gradually became extinct. The protons were initially massive—up to a million times their present size. There is no mechanism to identify the magnitude of their charge in relation to their mass. Thus the electron and the proton, objects of seriously different mass, have charges which although opposite in sign yet are of equal magnitude. If the charge of these newly formed giant protons was the same as it is nowadays for smaller protons, then charge repulsion would have been less effective as the massive protons collided with each other.

One significant effect was that the protons during their formation absorbed large amounts of gravitational energy (and Hubble acceleration energy) that was to be very slowly dribbled out over the lifetime of the protons. That acquired and stored energy was temporarily ineffective. This had the effect reducing the density of free gravitons, reducing the gravitational pressure and so allowing the time quanta to become more effective, resulting in the quickening of the velocities of the moving energy carrying particles. The velocity of the electrons and nucleons quickened. With their increased mass and higher velocity when the protons collided, the energy, to which was added their stored energy, was substantial, and this would have slowed the pace of time until such time as that energy was used, either for the work involved in the synthesis of a new nucleon or into the mass of the nucleon itself.

The direction of movement of the protons and antiprotons had to be random. They were concentrated and confined within the region of the surface of the decaying black hole and the curvature horizon. Collisions were inevitable. For the proton-antiproton collisions, they were aided and abetted by charge attraction. Just like the electron-positron debacle, there was an excess of positively charged protons, balancing the excess of negatively charged electrons so that gradually the antiprotons were annihilated.

But yet after the extinction of the antiprotons, the protons also continued to collide with each other. Charge repulsion was ineffective; the protons were too massive. Collision meant disintegration. The consequence was that protons disintegrated back to their original component parts. Their stored energy was released as a puff of gravitational energy, transiently slowing time around the debris, whilst the acceleration due to Hubble acceleration energy was brought to a halt. Then came the slow business of reassembling the various components, such as the various quarks, into the right order for the protons to be re-formed. The effect of that was any newly formed proton from the remnants of a proton-to-proton collision would have been smaller in mass. This must have occurred a number of times, but each time the newly re-formed nucleon was smaller in mass until such time as its mass was insufficient to overcome charge repulsion when two nucleons did collide.

This process of chaotic movement, collision, disintegration, and re-forming the nucleons in an ever expanding universe and slowing of time, with the cycle being repeated again and again, gradually led to the emergence of a large universe with an orderly outwards movement of the nucleons which were destined to form galaxies. The process was an extremely slow one, taking over a possible 100 billion years of cosmological time. The outwards movement of the nucleons was facilitated by the Hubble expansion energy, with an acceleration pace of 5×10^{-10} metres per second per second; that is, initially the velocity was very slow, although the pace of time was, at a minimum, 10 times, and possibly 100 times, faster than our present Earth time. All forces were relatively faster but were constant relative to the local pace of time.

Because of the high density of time quanta, the protons and neutrons became enormously large. That is, their energy levels were approximately a million times greater than their present values. The pace of time was also faster, a million times faster. This in turn meant that the speed of light was a million times faster. But as the proto-universe expanded, the density of the time quanta fell exponentially and with it the pace of time and so also the mass of the protons and neutrons. Every halving of the density (that is, doubling of the volume of the proto-universe) caused the nucleons to lose half of their mass as energy. Including in that energy release was gravity. That is, the nucleons aggregated together to form great clouds which pulled apart.

Now the evolution of the universe could move to a new phase. The great clouds were then able to move away at speeds well in excess of the present speed of light. The great clouds condensed into galaxies filled with giant massive stars. With the fast pace of time and huge masses, these giant stars generated huge quantities of heat. Hydrogen fusion controlled the temperature, but at the core of these huge stars, the local supply of hydrogen was insufficient and temperatures rose to the point of helium fission. The result was many giant supernovas blowing off their outer layers, which in turn formed smaller stars. Each giant star was left with a black hole residue, as the explosion violently compressed the very centre which had become packed with helium nuclei. The compression forced the electrons previously mixed around the naked helium nuclei into the helium nuclei, causing them to form neutrons. Thus very large black holes were formed. But the neutrons carried with them time quanta and the Hubble acceleration particles. That is, these black holes would move with the galaxy in which they were embedded; they also were gravitationally attracted towards each other to collide gently, because around them time was much slowed. This led to the formation of the giant black holes that are to be found in the centre of all spiral galaxies. With very large proto-galaxies, there could be two or more of these giant supermassive black holes, resulting in some pulling apart. These are the barred spiral galaxies.

But the galaxies were being accelerated by the Hubble acceleration particles so that within three doubling periods of cosmological time (slightly under 28 billion years), they had reached speeds approaching the speed of light. Meantime, the whole galaxy had expended much of its mass as energy, providing gravitational energy as well as electromagnetic radiation as a result of progressive time slowing, leaving behind an emaciated universe which was filled with photons and gravitons as well as the carriers for the other primary forces.

There is, though, a problem: ageing. Starting from scratch, the Hubble acceleration force will accelerate a nucleon up to the speed of light in 19 billion years if there is no interference. But there is: it is special relativity. As the velocity approaches the speed of light, special relativity slows down the pace of time in the moving object and slows down its release of other forms of energy such as electromagnetic energy, thus conserving mass. More importantly as far the mass is concerned, its velocity is apparently slowing, although relative to an independent observer in outside space, its velocity is still accelerating towards the speed of light, only the acceleration is slower. If the life of a nucleon (that is, a molecule, a star, a galaxy, a universe) is 30 billion years (this is being generous) and the age of the universe is approximately 169 cosmological years, how can this be resolved? There are two alternative hypotheses.

The first hypothesis is prolonged incubation. Time effectively started when the expansion of the primordial black hole had expanded to a radius of approximately half of our galaxy's radius. At this point the gravitational strength of the primordial black hole had become so diluted that it could no longer maintain an event horizon, with the result that the other forces could now escape. But the gravitational strength was still strong enough to maintain a tightly curved space, so that the escaping energy particles were confined to a rather narrow gap. More importantly, moving into this gap were the time quanta and heat photons. This raised the temperature enough for the time quanta to become effective. Photons of various kinds could race around the dying primordial black hole at great speed. The high density of the time quanta raised the pace of time until it was a million times faster than the present.

The primordial black hole still continued to disgorge its gravitons, allowing the curved space gradually to unroll, widening the gap. The whole zone containing the energy particles could now enlarge, and that gradually diluted the time quanta. This slowed the pace of time, and the swollen proto-nucleons shed some of their mass as energy as they slowed down. Eventually they became sufficiently small so that their mass and their inertial momenta were insufficient to overcome charge repulsion. Now they could disperse to form great clouds of nucleons and go on to form galaxies. But the whole thing must have taken at least 100 billion years before orderly expansion could take place.

The second hypothesis is recurring generation. It is possible that also present in space were (and possible still are) smaller primordial black holes. Their development had been arrested when hit by the huge amount

Chapter Eight

of energy released in that first doubling period. The amount of mass at the start was a million times greater than the present mass of the universe, and in that first doubling period, half the mass was lost as energy as the pace of time slowed to half speed.

With the near death of the first generation of galaxies, these primordial black holes would repeat the process that led to the formation of galaxies. Only this time the density of time quanta was much less (1/8th less) so that the protons and nucleons formed were smaller. Nevertheless, the same pattern occurred with the Hubble acceleration particles, causing them to accelerate to the point of destruction. And then another cycle of the generation of galaxies and stars could form. But with each generation, the size of the nucleons, the stars, the velocity of light, the pace of time itself was 1/8th that of the previous generation.

Thus the whole pattern of cosmology was the repetitive cycle of the equivalent of the observable universe forming, expanding, and dying only to be replaced by a succeeding generation of galaxies and stars, until we come to our present situation. This will continue until the entire stock of primordial black holes has been used up. Then the universe will consist of a vast void containing nothing but photons of various types, with time proceeding at an infinitesimally slow pace.

It is emphasised that this concept of a recurring visible universe is a hypothesis, but it does take into account all the pointers listed above and that are necessary for any valid system of cosmology. It rests on the possibility that underlies the duplication (or triplication) of the image of distant galaxies is responsible. Whatever is doing this is very large and very opaque. It cannot be a dead galaxy, as galaxies are mostly empty space and the gravitational force at the edge of a galaxy is incapable of producing acceleration as high as 5×10^{-10} metres per second per second, which raises the possibility that whatever is responsible for these duplications might be quiescent primordial black holes.

In summary, this proposal is that there are in the universe a number of primordial black holes. They all were heading for collapse and releasing the energy to form a universe. One particular one was first. Its energy inhibited the other primordial black holes from releasing their energy.

The newly formed universe could only last about 28 billion years of cosmological time (because of Hubble acceleration). When this one had gone, the inhibition of energy release from another primordial black hole was lifted, and so the race began for another primordial black hole to release its energy and repeat the cycle. This repetition must have occurred at least five times.

These two hypotheses are just that, hypotheses, logical hypotheses. There is insufficient evidence to say which is right or indeed that there might be another entirely different cosmology responsible for our present visible universe. But whatever it is, it has to take into account the Hubble acceleration force.

APPENDIX 8

A Partial Energy Budget per Kilogram of Helium Mass for Three Doubling Periods

Three doubling periods equal 26.5 billion years or 8.37×10^{17} sec. In that time, mass will have been lost as energy. Three doubling periods means that the pace of time will have halved three times; that is, mass must have been reduced to 1/8th of its original value. If the starting mass was 1 kg of helium, the residual mass would be 125 g. The mass will have decreased in an exponential fashion; that is, the mean mass for this period of time would be 353.6 g. In that time, gravitational energy will be shed at an average rate of 0.017 joules/sec/kg. The total energy shed as gravitational energy would therefore be 5×10^{15} joules.

The mass will also have been subject to acceleration according to the Hubble constant. This has a value of 5×10^{-10} m/sec/sec. The energy required for the acceleration of this average mass and the time taken (8.37×10^{17} sec) can be calculated as:

Equation Appendix 8.1 $e = mc^2/2$
$v^2 = (\text{acceleration} \times \text{time})^2$
Energy $= 3 \times 10^{16}$ joules

Together they amount to 3.5×10^{16} joules. From Einstein's famous mass energy equation, the amount of mass lost (875 g) amounts to 7.6×10^{16} joules. That is, gravity and acceleration energy account for slightly less than one half of the mass lost during these three doubling periods. The remainder of the loss is from the other sources of energy within the atoms of the initial mass—heat, electromagnetic energy, and so on. It has been suggested that at the very beginning, all the primary forces of nature were equal. Given that there are six primary forces of nature—the weak and strong intranuclear forces, electromagnetic force, gravity, the Hubble acceleration force, and the gravito-magnetic force—these crude figures suggest that that proposition may have some validity. This must apply to all masses, such as galaxies, stars, black holes, and so forth.

The Curvature Horizon

At the surface of the sun, the gravitational force is 272 m/sec/sec. This force is the result of the sun's gravitational force passing a distance of approximately 2.32 light seconds from the centre of the sun's mass. It causes space to curve by 2 arc seconds.

At the surface of the Earth, the gravitational force is approximately 10 m/sec/sec. There is no evidence showing that on Earth there is any curvature of a light beam; that is, there is, to all intents and purposes, no curvature of space around the Earth. It follows that when the gravitational force from a black hole falls to 10 m/sec/sec, any curvature of space produced by the black hole's gravitational strength effectively ceases. This is the curvature horizon. The inverse square law therefore predicts that:

Equation Appendix 8.2   Curvature horizon = Event horizon $\times (3 \times 10^8/10)^{0.5}$

That is, the curvature horizon for a black hole is approximately 5,500 times greater than its event horizon. For the massive black hole at the centre of our galaxy, 4 million solar masses, the curvature horizon is about 30 light hours. Given the size of our galaxy, this means much less than 1 billionth of our galaxy has any curved space. (The curvature is produced by all the billions of stars in our galaxy, most of which are G stars like our sun; their contribution to any curvature of space is negligible.) When to this is added the average intergalactic distance of empty space, the universe is flat, to less than a billion billionth of a per cent of the volume of the visible universe.

The sun, like all massive objects, also has a curvature horizon. Given that the gravitational acceleration at the surface of the sun is 272 m/sec/sec, compared with Earth's 10 m/sec/sec, the curvature horizon is $27.2^{0.5}$ times the sun's radius. It follows that light from distant stars that passes approximately 5 solar radii from the surface of the sun will show no apparent displacement during a solar eclipse. This is testable.

CHAPTER NINE

Lists of Findings and Proofs

This book arose out of an interest in trying to understand the physics underlying the Hubble constant. This unexpectedly led to a whole series of new discoveries. The list is as follows:

Chapter 1

1. The Hubble constant is, after allowing for relativity effects, constant to the edge of the visible universe.

2. The Hubble constant has a value of 52 +/-1 km/sec/Mpc.

3. The age of the universe in Earth time units is 12.5 billion years. This figure arises from an extrapolation of the present pace of time and assumes that the period of the second is and has always been immutable.

4. At very high velocities, the special relativity effect dims the light of Type 1A supernovas, leading to an overestimation of their distances and so an underestimate of the Hubble constant.

5. At very high velocities, the special relativity effect interferes with the red shift of the receding supernova, causing a small underestimate of its velocity.

6. The Hubble constant is describing an acceleration constant which has a value of 5×10^{-10} m/sec/sec.

7. This constant is a fundamental constant of nature whose existence has been previously proposed but hitherto not found or quantified.

8. The acceleration force arises from within the nucleons, just as gravity does.

9. Without this acceleration force, the universe would have collapsed into a gravitationally bound compressed mass of matter. That is, this acceleration force is responsible for the expansion of the universe.

10. Dark matter does not exist.

11. Dark energy does not exist.

12. Time within a black hole is normal.

Chapter 2

13. The Hubble constant can only be constant if time is expanding (slowing).

14. The distances to galaxies on the opposite sides of the universe show that the universe must be much older than 14 billion years, as the sum of these distances cannot exceed the age of the universe.

15. The period of the second is not constant.

16. All the frequencies of the colour spectrum that make white light are reduced when the source of the light is receding at high velocity, causing some red light wavelengths to be stretched to become waves signalling infrared.

17. When this occurs, white light is only maintained by the recruitment of ultraviolet light, whose waves are stretched to become that of violet, so maintaining the white colour balance.

18. At near luminal speeds of the light emitter, this compensatory red drift starts to fail and the red light frequency begins to dominate, tinting the white light.

19. The frequency composition on photons cannot change, once the photons are in transit.

20. The frequency composition of light from a distant moving source, including any red shifted frequencies, is governed by the speed of light which in turn is governed by the environmental pace of time through which the light is passing.

Chapter 3

21. The universe has its own system of time (cosmological time) whereby time, the period of the second, has been expanding exponentially.

22. Every time the universe doubled its Earth time age, the pace of time, or period of the second, doubled. This is a continuous process.

23. Each doubling episode has the same quantity of cosmological time, viz. 8.84 billion years of cosmological time.

24. The period of a cosmological second has converged to become that of an Earth time second to within less than a billion billionth of a second.

25. The ratio of ages that is the Earth time age of the universe at any past event divided by the present Earth time age determines the extent of time expansion since that past event.

26. Time and mass are inversely related, so that doubling the period of the second halves mass.

27. The speed of light adheres to the prevailing pace of time.

28. The period of the second has doubled 19.1 times since the beginning of time.

29. When time started the period of the second was 1/650,000th its present value.

30. The age of the universe in cosmological time is 169 billion years.

31. The universe adheres to cosmological time and not Earth time.

32. The universe started from a primordial black hole that had a radius of 23,000 light years.

33. The Big Bang theory does not explain the origin of the universe.

34. The lifetime of any recognisable galaxy is 20 to 25 billion years.

35. There has been a considerable turnover of galaxies, with the earlier ones accelerated to the speed of light, turning into dust as they do so.

36. The Schwartzchild equation, defining the event horizon of a black hole, is inappropriate to conditions occurring at the origin of the universe.

Chapter 4

37. The sun started with a dowry of 10 per cent helium.

38. Hydrogen fusion is endothermic.

39. Time slowing causes the sun to lose $\sim 3 \times 10^{11}$ kg of mass per second.

40. Time slowing, through its effect on mass, is responsible for the sun's radiant energy output.

Chapter Nine

41. All nuclear fusion is endothermic, unless the sum of the atomic numbers of the fusing elements is greater than the atomic number of the element produced by atomic fusion.

42. The sun's radiant energy output is 4.4×10^9 kge/sec, where 1 kge is equal to 9×10^{16} joules.

43. The sun synthesises 1.2×10^{12} kg of helium per second from hydrogen fusion.

44. Radiant energy accounts for ~1 per cent of the energy released from the sun by time expansion.

45. Hydrogen fusion is endothermic and in forming helium absorbs approximately ~2 per cent of the energy released by time slowing.

46. Solar gravity accounts for approximately 97 per cent of the energy released by time slowing.

47. The remaining life of the sun is less than 15 billion years.

48. The sun expends nearly 2000 kge of energy per second to maintain the planets in their orbits.

49. The gravitational energy used to maintain the orbits of Venus and Earth is greater than the amount of gravitational energy those planets intercept directly from the sun.

50. Helium fission requires a temperature of around 30 million degrees Celsius.

Chapter 5

51. Gravity is not a consequence of any curvature of space.

52. Gravity causes some refraction of light.

Chapter Nine 134

53. At a distance of 5 solar radii from the sun, there can be no detectable curvature of a ray of light caused by the sun.

54. The wave/particle dilemma of photons and gravitons is due to the lateral spreading of the energy that makes the particle. This spreading can be retracted, so reducing the wave width if the wave has to pass through a narrow gap.

55. Gravity works by interacting with Hubble accelerons (responsible for the Hubble constant), inhibiting them on the proximal surface of the nucleon, then during the graviton's transition through the nucleus of the nucleon, changing to accelerons.

56. Space is full of gravitons which criss-cross each other so that there is no region in space which is free of gravitons. The further they travel, the wider their wave width.

57. If some gravitons are absorbed by nucleons, the adjacent gravitons will expand and move sideways to fill the gap in the overall wave front. This can include gravitons that are outside the mass. This accounts for why the gravitational acceleration of an irregular mass is independent on the orientation of that mass, and why gravity casts no shadow.

58. The Newton is constant in all time frames.

59. Mass is constant in all time frames.

60. Mass passing through to or experiencing a slower time frame, or pace of time, will shed mass as energy.

61. The universal slowing of time that is part of the expansion of the universe through its effect on mass causes the release of the fuel for gravitational and radiation energy.

62. It follows that gravity is constant relative to its time frame. This forms the base for the fundamental constant known as the gravitational constant.

Chapter Nine

63. If the gravitational constant is constant in all time frames, it follows that length, the metre, is also constant in all time frames. The faster the pace of time, the longer the metre. Conversely, the more expanded the time, the smaller the metre. This only applies to the moving object.

64. If distance is measured or calculated using light seconds or light years, any changes in the pace of time experienced by the observer whilst travelling along that distance will not affect that distance.

65. Wavelength is also affected by the pace of time. The faster the pace of time, the longer the wavelength.

66. Frequency of any radiant wave is unaffected by any change in the pace of time; its equivalent, the velocity of light, is constant in all time frames. As a consequence the light and spectra of light from a very distant galaxy or supernova where time was faster will be unchanged during transit—there is no stretching of the waves.

67. The Doppler or red shift reflects the velocity of the light emitting mass relative to the pace of time of that mass.

68. Einstein's conclusion that at high velocities mass increases alongside the expansion of time is incompatible with the principle that the laws of physics are the same for all observers.

69. Recalculation of relativity's mass velocity equation, based on the law of conservation of momentum, shows that at high velocities mass decreases to the same extent as time expands.

70. Gravitational energy is released at a rate of 0.0176 joules/sec/kg.

71. The event horizon of a black hole is about one third of the wavelength of a graviton above the surface of the black hole.

Chapter 6

72. The moon arose from the impact of an asteroid striking the Earth.

73. The Earth lost 1.23 per cent of its mass, unbalancing the Earth's rotation.

74. The impact generated tsunamis over 15 miles high, which swept around the globe.

75. The crater was as big as the combined sizes of the Pacific and Atlantic Oceans.

76. The crater was over 300 km deep.

77. All land levels fell by over 30 km.

78. The loss of heat generated upon that impact created Snowball Earth, which lasted over 2 billion years.

79. The slow restoration of internal heat melted Snowball Earth, thinned the Earth's cold thickened crust, and so enabled the delayed plate tectonics to develop to restore the balance of the Earth's rotation. In the west, plate tectonics did not develop until after the start of the evolution of primates.

80. The evolution of life itself was delayed by Snowball Earth.

81. Snowball Earth protected the Earth's surface from impacts from the great meteorite storm that pockmarked the moon.

82. The Earth's atmospheric oxygen came from subduction of sand and sandstone into the magma, which caused the silicon atoms in the sand molecules to dissociate, releasing the oxygen which eventually escaped through volcanic action.

83. The excess heat production from the Earth is due to time slowing releasing thermal energy.

84. The universe's time slowing predicts the peak high tide around various islands in the Pacific Ocean to within 4 per cent of their mean observed value.

85. The background microwave radiation cannot come from any sort of Big Bang.

86. The solar system must be surrounded by a shell of low density dust at a distance equal to or just beyond the Oort clouds.

87. That dust is releasing heat energy at an extremely low rate due to the effect of the expansion of time on mass.

88. The hot spots seen in the microwave radiation are due to random patches of dust that have a slightly higher density than other parts of the dust shell.

Chapter 7

89. Jupiter's equatorial belt of very energetic storms are the result of time slowing at the equator consequent upon the planet's high rate of revolutions and its giant size.

90. The breakup of comet Shoemaker-Levy was a consequence of the build-up of steam within the comet plus the loss of surface ice as a result of time slowing due to high velocity caused by Jupiter's gravitational acceleration.

91. Gravitational compression of giant gas clouds could concentrate the heat so that hydrogen fusion temperature is reached, but it requires rather specific starting conditions that render it highly improbable that gravity is the main cause of stars reaching luminous temperatures.

92. Jupiter's moon Io's active volcanoes must be the consequence of the energy released by radioactive decay.

Chapter Nine

93. Saturn's moon Enceladus's ice fountains are the most spectacular sight in the whole solar system. If the ice fountains were on Earth, they would reach a height of several kilometres.

94. Saturn's ring system is the result of the out-jetting of ice fountains and then collapse and fragmentation of other pure ice moons similar to Enceladus.

Chapter 8

95. The Big Bang theory of the origin of the universe cannot be sustained.

96. Scattered throughout space are primordial black holes which cause multiple images of galaxies that are some distance behind them.

97. The age of our present visible universe, after allowing for relativity effects, is approximately 26 billion years due to the Hubble acceleration constant.

98. Our present visible universe will be replaced by galaxies that emerge from the present quiescent primeval black holes.

99. Time exists as quanta of time, just as its reciprocal, energy, exists as quanta of energy.

The Nature of Proof

All hypotheses require a measure of proof. Sometimes that proof can be obtained by experiment, that is, physically confirming the predictions made by the hypothesis. But in certain areas, such as biology and astronomy, experimental evidence is not available and the proof then lies in that the hypothesis provides a simple explanation for a wide variety of observations that are otherwise inexplicable. Darwin's theory of evolution is such a case.

Chapter Nine

Newton's theory that there was a force he called gravity provided a simple explanation for such things as the orbits of the planets, as well as why things fall to the ground, was so self-evident that no additional proof was required. Another type of proof is if the hypothesis leads to another hypothesis that can be tested. Einstein's special theory of relativity led directly to his mass energy equation, which was testable, and so the special theory came to be accepted as valid. It is sometimes said that extraordinary hypotheses require extraordinary proofs. This is simple discrimination against new ideas. The logic should be if the hypothesis provides explanations for events for which there is no known explanation within the proven laws of science, and there is no evidence to disprove the hypothesis, then it is reasonable to accept the hypothesis, whilst acknowledging that it could be wrong. Then there are the hypotheses which provide an explanation which breaks the proven laws of physics; the inflation theory of the origin of the universe comes into this category.

This study, the expansion of time, has only one direct example or proof and that is a mathematical one, to account for the constancy of the Hubble constant: but it led to the realisation that time and mass are inversely related and that led to a large number of proofs as well as a testable prediction.

The Proofs

1. The constancy of the Hubble constant. This constancy was shown to exist up to the limit of the observable evidence, that is Type 1A supernovas. Standard Euclidean mathematics predicted that it should not be constant unless adjustment was made for the expansion of time. There is no other mathematical possibility.

2. The Mid-Pacific tides. The hypothesis predicted the amount of gravitational energy released by the moon and the amount of that intercepted by the Earth, and so consequently the amount of energy available for the change in height of the tides in the middle of the Pacific Ocean. The prediction matched the facts to a surprising degree of accuracy when one considers that the predictions rested purely upon knowing the mass of the moon and

its average distance from Earth and the calculation of the age of the Earth in cosmological time.

3. The heat of the gas planets. All the gas giants are much warmer than their environment in space, for which there is no explanation. Further, as was shown by the Jupiter landing probe, the internal heat was even greater than the surface temperature. So much heat energy requires some form of mass-to-energy conversion, and the only available explanation is time expansion and its associated inverse relationship with mass.

4. The stability of G stars. Stars like are the sun are very stable for millions if not billions of years. Yet they are losing large amounts of mass as energy every second. It has been maintained that this is because of a balance between the heat output and its consequent expansion effect and the star's (or the sun's) gravitational force. Hydrogen fusion, if it was exothermic, would not be able to maintain such a steady balance over millions of years for all the billions of G stars that exist. Indeed it would result in the sun's rapid demise as a supernova. The only explanation for such long lasting stability is that hydrogen fusion must be endothermic, and this acts like an automatic thermostat. A corollary, not discussed earlier, is that more massive stars produce much more heat energy. Despite their greater size and therefore the greater amount of hydrogen, they consumed their hydrogen reserves more quickly and so ended up with a much hotter surface temperature. It would not take long before they reach helium fission temperature, when they would erupt as giant supernovas.

5. The background microwave radiation. The hypothesis predicts that if the solar system is surrounded by a layer of thin dust, then because of time slowing that dust will gradually lose energy as heat that is microwave radiation. The amount of heat is minute but because it surrounds the solar system as well as radiating outwards, it would be radiating inwards and the nearer we are to the centre of the solar system, more energy will be concentrated and so detected. Furthermore, the intensity of the radiation, the amplitude of its waves, and the narrowness of the bandwidth

point to the fact that the source surrounds us and is close by. Fundamental to this conclusion is that electromagnetic waves cannot change their frequency whilst in transit, although they can change their amplitude (through the inverse square law). It has been postulated that they are the remnant of the Big Bang. But that would have been at a very high temperature and therefore caused high frequencies in the radiation. These are not seen. The hot spots seen in the microwave distribution are exactly what one would expect if the layer of dust was irregular. Observation of the Crab nebula shows that a supernova does create a rim of dust.

6. Enceladus's icy fountains. Given that this little moon is almost pure ice, the only mechanism that could produce such an extraordinary scene as the huge ice fountains is the pressure from superheated steam erupting into the intense cold of space. The amount of energy required is immense. This, added to the loss of heat from the surface of this little moon, means that the energy can only come from some form of mass energy conversion. The expansion of time through its effect on mass is the only explanation available. It also explains Saturn's majestic ring system.

Chapter Nine

The Three Paradoxes

1. **The distance paradox 1.** The sum of the distances in light years of two galaxies that lie on opposite sides of the visible universe cannot exceed the age of the visible universe in years. It would take time for material to get that far, form stars and galaxies, the stars then go through their life cycles before sending the light from their supernovas to travel back to Earth. The Sloane survey shows that we are surrounded by galaxies at least 10 billion light years distance away. It follows that the age of the visible universe must be in excess of 13 to 14 billion Earth time years. The expansion of time, and that the universe uses cosmological time, resolves this paradox.

2. **The distance paradox 2.** The latest telescopes show light that is postulated as occurring very soon after the Big Bang. That light has been travelling for well in excess of 12 billion years. At that time, what is now the visible universe was very relatively small. That is, that light would not have to have travelled in excess of 10 billion years. That is, the visible universe must be much older than the postulated 13.7 billion years. The cosmological age, which relies on expanding time, eliminates this problem.

3. **The globular cluster paradox.** Many of the 150 globular clusters found within our own galaxy have been shown to be older than 14 billion years. This is impossible if the visible universe is 13.7 billion years old. The clusters themselves would have taken several billion years to have formed if they are the result of a supernova explosion of a giant star. That is, our galaxy must be considerably older than 13.7 billion years.

www.ingramcontent.com/pod-product-compliance
Lightning Source LLC
Chambersburg PA
CBHW030749180526

45163CB00003B/957